Marine Ecosystems - Biodiversity, Ecosystem Services and Human Impacts

Edited by Ana Marta Gonçalves

Published in London, United Kingdom

Marine Ecosystems - Biodiversity, Ecosystem Services and Human Impacts
http://dx.doi.org/10.5772/intechopen.95153
Edited by Ana Marta Gonçalves

Contributors
Abhijit Mallik, Suchismita Prust, Puja Chakraborty, Shyamal Chandra Sukla Das, Shashi Bhushan, Adib Ali Saad, Lana Ezaldin Khrema, Melekote Nagabhushan Arun, Rapolu Mahender Kumar, Banugu Sreedevi, Guntupalli Padmavathi, Pallakonda Revathi, Neha Pathak, Dayyala Srinivas, Raman Meenakshi Sundaram, Sailaja Nori, Rajuddin Syamsuddin, Ana Marta Gonçalves, Andreia Filipa Mesquita, Fernando José Mendes Gonçalves, Neila Annabi-Trabelsi, Mohammad Ali, Genuario Belmonte, Habib Ayadi, Wassim Guermazi

First published in London, United Kingdom, 2024 by IntechOpen
IntechOpen is the global imprint of INTECHOPEN LIMITED, registered in England and Wales, registration number: 11086078, 5 Princes Gate Court, London, SW7 2QJ, United Kingdom
Printed in Croatia

British Library Cataloguing-in-Publication Data
A catalogue record for this book is available from the British Library

Additional hard and PDF copies can be obtained from orders@intechopen.com

Marine Ecosystems - Biodiversity, Ecosystem Services and Human Impacts
Edited by Ana Marta Gonçalves
p. cm.

This title is part of the Environmental Sciences Book Series, Volume 15
Topic: Ecosystems and Biodiversity
Series Editor: J. Kevin Summers
Topic Editors: Salustiano Mato, Josefina Garrido and Francisco Ramil

Print ISBN 978-1-83968-459-3
Online ISBN 978-1-83968-460-9
eBook (PDF) ISBN 978-1-83968-544-6
ISSN 2754-6713

For EU product safety concerns:
IN TECH d.o.o., Prolaz Marije Krucifikse Kozulić 3, 51000 Rijeka, Croatia,
info@intechopen.com or visit our website at intechopen.com.

IntechOpen Book Series
Environmental Sciences
Volume 15

Aims and Scope of the Series

Scientists have long researched to understand the environment and man's place in it. The search for this knowledge grows in importance as rapid increases in population and economic development intensify humans' stresses on ecosystems. Fortunately, rapid increases in multiple scientific areas are advancing our understanding of environmental sciences. Breakthroughs in computing, molecular biology, ecology, and sustainability science are enhancing our ability to utilize environmental sciences to address real-world problems.

The four topics of this book series - Pollution; Environmental Resilience and Management; Ecosystems and Biodiversity; and Water Science - will address important areas of advancement in the environmental sciences. They will represent an excellent initial grouping of published works on these critical topics.

Meet the Series Editor

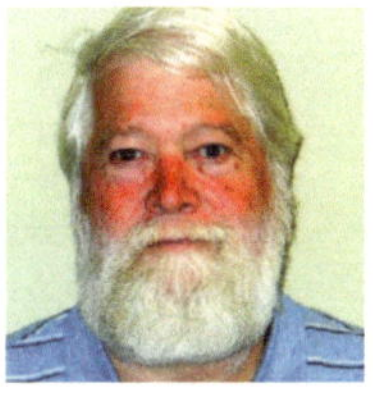

J. Kevin Summers is a Senior Research Ecologist at the Environmental Protection Agency's (EPA) Gulf Ecosystem Measurement and Modeling Division. He is currently working with colleagues in the Sustainable and Healthy Communities Program to develop an index of community resilience to natural hazards, an index of human well-being that can be linked to changes in the ecosystem, social and economic services, and a community sustainability tool for communities with populations under 40,000. He leads research efforts for indicator and indices development. Dr. Summers is a systems ecologist and began his career at the EPA in 1989 and has worked in various programs and capacities. This includes leading the National Coastal Assessment in collaboration with the Office of Water which culminated in the award-winning National Coastal Condition Report series (four volumes between 2001 and 2012), and which integrates water quality, sediment quality, habitat, and biological data to assess the ecosystem condition of the United States estuaries. He was acting National Program Director for Ecology for the EPA between 2004 and 2006. He has authored approximately 150 peer-reviewed journal articles, book chapters, and reports and has received many awards for technical accomplishments from the EPA and from outside of the agency. Dr. Summers holds a BA in Zoology and Psychology, an MA in Ecology, and Ph.D. in Systems Ecology/Biology.

Meet the Volume Editor

Dr. Ana Marta Gonçalves obtained a Ph.D. in Biology with a specialty in Ecology from the University of Coimbra, Portugal, in collaboration with Ghent University, Belgium, in 2011. Since 2019, she has been an auxiliary researcher at the Department of Life Sciences, Faculty of Sciences and Technology, University of Coimbra. During her research career, Dr. Gonçalves obtained several grants in highly international competitive calls, including the 2008 MARS award for young scientists funded by the Royal Netherlands Institute for Sea Research (NIOZ) and the Foundation for Science and Technology (FCT, Portugal) grants, and an honorable mention for the Impact of International Publications in 2022, according to Web of Science, by the Director of the Faculty of Sciences and Technology, University of Coimbra. She was a member of the board of directors of the research unit Marine and Environmental Sciences Centre (MARE). Currently, she is the chair of the *Society of Environmental Toxicology and Chemistry* (SETAC) Europe Education Committee, coordinator of LTsER Estuaries platform, eLTER coordinator of the Working Group Sites and Platform Forum (SPF) Training Group, member of the board of directors of IMAR – Institute of Sea, an associate member (representative) of the EuroMarine and IMBRSea networks, and a coordinator of La Caixa Foundation fellowship programs. Since 2016, Dr. Gonçalves has been a member of the Scientific Council of Institute for Interdisciplinary Research, University of Coimbra. She is a journal editor and reviewer and has edited one book. Dr. Gonçalves is an advisory board member for Food Sciences and Technology and Toxicology. She is also a coordinator and research member of several national and international projects. She holds various administrative and science management positions in international networks. Dr. Gonçalves has supervised more than 140 graduate and post-graduate students and published more than 114 indexed papers and more than 55 book chapters. Dr. Gonçalves develops her work in the field of biological sciences, in the frontier of aquatic ecology and ecotoxicology, and biochemical pathways, assessing potential changes in trophic food webs, from primary producers to top predators, exposed to chemical stressors and climatic changes. She also researches marine biotechnology and the valorization of marine resources with industrial applications like food, biomedicine, pharmaceutical and agriculture novel products. Dr. Gonçalves dedicates part of her time to health and ocean literacy and environmental education. Her areas of specialization include marine biotechnology aquatic ecotoxicology, trophic food webs, global changes, anthropogenic impacts, microplastic pollution, biochemical analysis (fatty acid and carbohydrate profiles, proteins), biomarkers, bioindicators, aquatic ecology, estuarine plankton dynamics, long-term changes of plankton communities, aquaculture, and environmental education.

Contents

Preface

Marine ecosystems are among the most populated habitats in the world. Their richness is given by an incredible number of marine organisms that provide human beings with ecosystem services. The Millennium Ecosystem Assessment defined "ecosystem services" as "the benefits that people obtain from ecosystems." Marine biota is a great source of ecosystem services.

Ecosystem services are dependent on natural ecosystem functions, thus it is important to preserve our planet and its habitats. For example, the provision of seafood is the most exploited ecosystem service, which is a result of a combination of primary and secondary production, biogeochemical cycling, food web dynamics, and so on. The economic benefits are also encouraging; it has been estimated that marine ecosystem services are valued at $50 trillion per year. Marine species mostly used for human benefits are principally known in the food industry, but they can also assure several benefits in different biotechnological applications, such as in the nutraceutical, pharmaceutical, or cosmetics industries, as demonstrated by studies on seaweeds and microalgae. The interest in and search for new natural compounds from marine biodiversity to discover novel bioactive substances has increased in the last decade. It is expected that many naturally bioactive marine compounds with novel structures and bioactivities may be found among marine metabolites. Still, with the constant and aggravated depletion of our natural environments, such as supplies of water and soils, the foundations of our food production are compromised. In this scenario, there is a great interest in marine organisms such as macroalgae biomass and resource exploration, with a substantial body of evidence supporting the nutritional potential of macroalgae. Marine macroalgae produce significant amounts of primary and secondary metabolites that present a wide variety of bioactive properties, including antioxidant, antiviral, antimicrobial, antitumoral, anti-inflammatory, anti-aging, or immunomodulatory potential, along with antibiotic properties. These molecules are promising candidates for many possible applications such as in pharmaceutical, nutraceutical, cosmetics, and medical industries as well as agriculture and feeding. These applications may benefit the human quality of life. Among benthonic organisms, mussels are also used in different sectors, for example, the anti-inflammatory agent contained in Lyprinol may be extracted from a green-lipped mussel originating from New Zealand. Moreover, sponges and coral are also rich in bioactive compounds that could be useful for future studies and the development of new human services. Indeed, ecosystem services offered by marine resources are diverse. For example, mangrove ecosystems are important to ensure nursery and breeding habitats for many important fish species that can be exploited for aquaculture, providing other important services associated with habitat functions and protection. Connected with the extraction and harvesting of natural marine resources are activities related to tourism and culture Commercial fishing and marine tourism are both highly dependent on ecosystems, but marine tourism has been identified as having a lower impact on the environment and higher monetary value. Unfortunately, anthropogenic impact affects our ecosystems, leading to habitat degradation, pollution, overfishing, and more. Due to the unsustainable human exploitation of natural resources, ecosystem services are limited. Thus, to avoid the complete depletion of marine environments and to have functional ecosystem services over time, it is necessary to preserve and adopt sustainable uses of the environment. The Marine Strategy Framework Directive suggests the use of biological indicators to recognize stressed ecosystems to facilitate the process of restoration. The importance of zooplankton groups and their contribution to ecosystem services is apparent by examining the ecological roles of zooplankton and their impact on fish recruitment and fishery management.

This edited volume is a collection of reviewed and relevant research chapters concerning the developments within marine ecosystems. Chapters in this book include:

Chapter 1: "Evaluation of the Water Quality and the Eutrophication Risk in Mediterranean Sea Area: A Case Study of the Gulf of Gabès"

Chapter 2: "Biostimulant Properties of Marine Bioactive Extracts in Plants: Incrimination toward Sustainable Crop Production in Rice"

Chapter 3: "Seaweed *Kappaphycus alvarezii* Cultivation for Seagrass Ecosystem Conservation"

Chapter 4: "Marine Bivalves' Ecological Roles and Humans-Environmental Interactions to Achieve Sustainable Aquatic Ecosystems"

Chapter 5: "Non-Indigenous Marine Fish in Syria: Past, Present and Impact on Ecosystem, and Human Health"

Chapter 6: "Fish Otolith Microchemistry as a Biomarker of Metal Pollution in the Estuarine Ecosystem"

Ana Marta Gonçalves
Faculty of Science and Technology,
Department of Life Sciences,
University of Coimbra,
Coimbra, Portugal

Chapter 1

Evaluation of the Water Quality and the Eutrophication Risk in Mediterranean Sea Area: A Case Study of the Gulf of Gabès

Neila Annabi-Trabelsi, Mohammad Ali, Genuario Belmonte, Habib Ayadi and Wassim Guermazi

Abstract

The Gulf of Gabès, located in southern Tunisia, is a distinct and ecologically significant area in the Mediterranean Sea. Unfortunately, this dynamic marine ecosystem is experiencing cultural eutrophication, a process where water enrichment with nutrients like phosphorus and nitrogen salts leads to excessive algae growth, disrupting the ecological equilibrium and degrading water quality. In the Gulf of Gabès, key sources of nutrient pollution include industrial discharges, urbanization and agriculture. Eutrophication's effects here include harmful algal blooms, oxygen depletion, and declining water quality, upsetting the marine ecosystem's balance and impacting both fish and aquatic life. Nutrient enrichment interacts with trace metal pollution, overfishing and climate change. Future research must acknowledge and consider the complex interactions among these variables. Efforts in the Gulf of Gabès to address eutrophication involve tighter industrial regulations, enhanced agriculture and improved wastewater management, all crucial for preserving the marine environment's integrity and ensuring sustainability for the future.

Keywords: eutrophication, phosphorus, phytoplankton, zooplankton, algal blooms

1. Introduction

Cultural eutrophication poses a severe environmental and economic challenge in coastal marine ecosystems across the globe [1–3]. According to the European Union's definition, cultural eutrophication involves enriching water with nutrients, especially compounds of nitrogen and phosphorus, causing an accelerated growth of algae and higher forms of plant life to produce an undesirable disturbance to the balance of organisms present in the water and to the quality of the water concerned [4]. This kind of pollution by nutrients is resulting from anthropogenic activities including industry, agriculture and sewage disposal [5]. The repercussions of cultural eutrophication extend to coastal ecosystem biodiversity and the services they give to society such as protection from coastal erosion and flooding, or fish production [6].

Eutrophication is mostly accompanied by hypoxia and an increase in the biomass of nuisance algal taxa [7]. In fact, eutrophication is an important factor contributing to the increased frequency and diversity of harmful algal blooms [8] and shifts in the zooplankton community [7]. Despite the fact the Mediterranean Sea is generally considered of good quality [9, 10] and classified as nutrient-poor and oligotrophic [11], eutrophication problems began emerging in the 1960s [12], mainly situated in areas of encircled gulfs and bays near big cities, in estuarine areas and near ports [13–17]. The Gulf of Gabès, situated on the southeastern coast of Tunisia, faces pollution from anthropogenic and industrial activities, and its coastal area showed signs of eutrophication [18]; continuous and increasing deterioration of the coastal waters of the Gulf was reported since the industrialization [19–21].

2. The Gulf of Gabès: main characteristics

Situated in southeastern Tunisia, the Gulf of Gabès, also known as "Petite Syrte", stretches from 9.5 to 12°E in longitude and from 33 to 35.5°N in latitude. It extends from "Ras Kapoudia" in the north to the Tunisian–Libyan border in the south, encompassing a coastline exceeding 400 km (**Figure 1**). It shelters various islands (Kerkenneh, Kneiss and Jerba) and lagoons (Boughrara and El Bibane).

It is characterized by a semiarid Mediterranean climate, shallow waters, weak currents, high salinity and high temperature. The Gulf's circulation is predominantly influenced by tides and anticyclonic winds [22].

The Gulf has the highest tidal range in the Mediterranean Sea (maximum >2 m), essentially due to the low slope of the continental shelf and the shallow depth, which maintains its horizontal dimensions close to the resonance phenomena [22, 23].

Since industrialization in 1970, the Gulf of Gabès has been heavily impacted by industrial wastes and described as one of the most human-impacted coastal areas in the Mediterranean Sea [21]. Diverse untreated pollutants from liquid and solid wastes discharged from industrial and domestic activities (crude phosphate treatment, chemical processing plants, tanneries and textile mills) have severely degraded the Gulf of Gabès [24, 25]. The phosphate fertilizer industry is mediated as the major source of pollution. The most important phosphate industry group (Tunisian Chemical Group: TCG) was set in Gabès city. Two other companies (SIAPE and GRANUPHOS) are located in Sfax city, and recently, another one (SIAPE II) has been launched in Skhira. The unregulated disposal of phosphogypsum (PG), a byproduct of the phosphate fertilizer industry, into the Gulf of Gabès at a daily rate ranging from 1000 to 13,000 tons, stands as the primary factor responsible for the deterioration of this ecosystem [26, 27].

3. Causes of eutrophication in the Gulf of Gabès

Phosphogypsum (PG) industrial effluents released into the Gulf of Gabès are notably acidic and carry elevated levels of fluoride and phosphate. Additionally, they contain variable concentrations of heavy metals such as cadmium, chromium, copper, zinc and lead, along with radionuclides [28, 29]. The study of El Kateb et al. [30] clearly established that the substantial discharge of PG is a leading cause of eutrophication in the Gulf of Gabès.

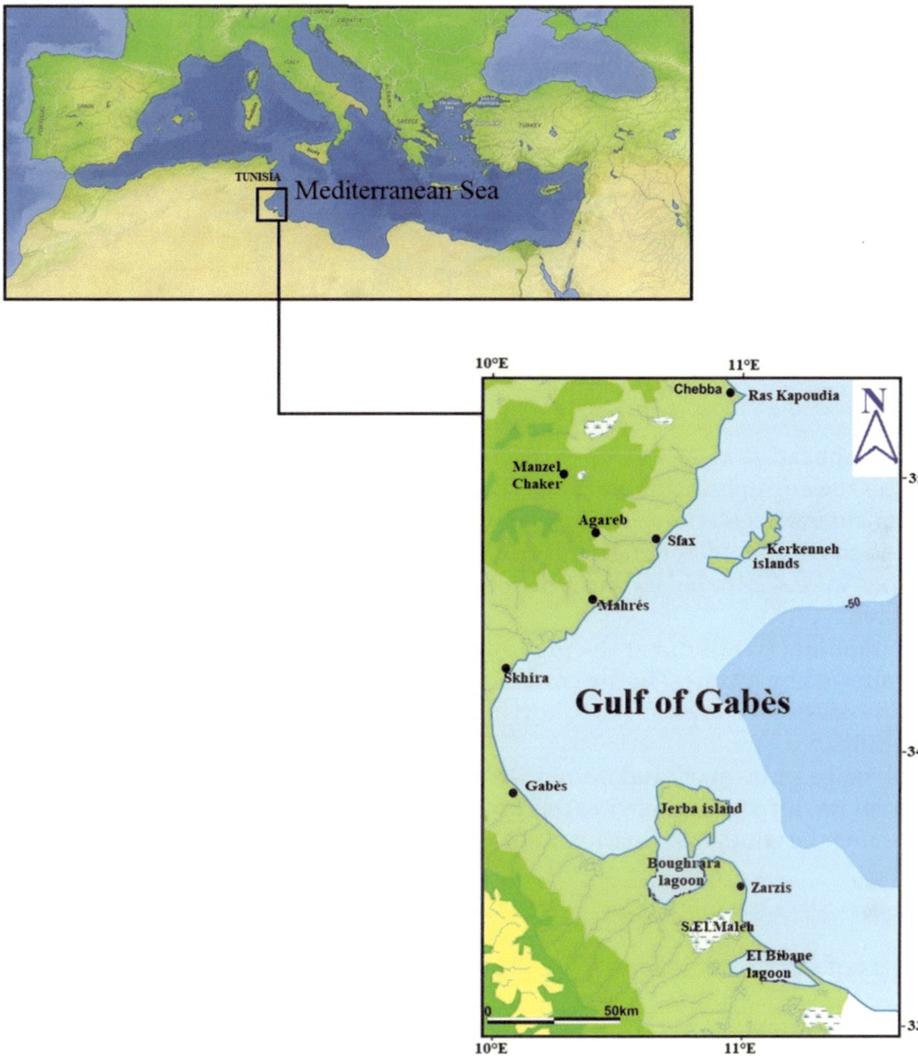

Figure 1.
Localization of the Gulf of Gabès.

While phosphorus is an essential element in the marine ecosystem, its excessive enrichment poses various challenges to biota, elevates the growth of organic matter, and consequently reinforces eutrophication [31, 32]. The high availability of inorganic phosphate along the coast of Gabès is also associated with agricultural use of land along the coast [33]. For phosphates, eutrophication is identified by phosphate levels exceeding 0.68 μmol l^{-1} [34]. Very high concentrations of phosphates are recorded in coastal areas of the Gulf of Gabès and reflect a eutrophic state. In fact, phosphate concentrations ranged from 0.51 to 8.52 μmol L^{-1} (mean ± SD = 2.98 ± 2.44 μmol L^{-1}) in the southern coast of Sfax [35]. They fluctuated between 1.25 to 2.97 μmol L^{-1} (mean ± SD = 2.07 ± 0.62 μmol L^{-1}) in the northern coastal area of Sfax [36]. In the coastal waters of Gabès, phosphate concentrations ranged from 1.7 to 6.7 μmol L^{-1} (mean ± SD = 3.74 ± 1.65 μmol L^{-1}) [17].

	Southern coast of Sfax [35]	**Northern coast of Sfax [36]**	**Coast of Gabès city [17]**
Ammonium (μmol L^{-1})	3.47–38.06 (13.26 ± 1.85)	2.99–7.75 (5.01 ± 1.71)	2.8–20.6 (8.73 ± 5.41)
Nitrates (μmol L^{-1})	1.09–26.53 (7.56 ± 1.5)	1.71–11.44 (3.07 ± 2.95)	3.1–6.0 (0.6 ± 0.1)
Nitrites (μmol L^{-1})	0.02–4.83 (0.82 ± 1.29)	0.04–2.64 (0.37 ± 0.8)	0.4–0.7 (3.07 ± 2.95)

Eutrophication is recognized for ammonium when values are above 2.2 μmol L^{-1} [34].
Eutrophication is recognized for nitrates when values are above 1.19 μmol L^{-1} [34].

Table 1.
Minimum, maximum and mean values of dissolved inorganic nitrogen forms in the Gulf of Gabès ± standard deviation from previous studies.

The application of an Eutrophication Index (E.I.) [37] was used by [17] in order to assess the eutrophication status of the coastal waters of Gabès. This index takes into consideration nitrites, nitrates, ammonium, phosphates and Chlorophyll-a. If E.I. is above 1.51, water quality is bad. Values of E.I. calculated in the Gulf of Gabès were > 2.16, indicating a high ecological alteration of the marine ecosystem [17].

Ammonium stands out as the predominant form of dissolved inorganic nitrogen, followed by nitrates, in the coastal waters of the Gulf of Gabès [17, 35, 36]. This dominance is a typical characteristic of untreated anthropogenic wastewater input [38].

The range (min–max) and mean values of dissolved inorganic nitrogen forms (ammonium, nitrates and nitrites) recorded in the Gulf of Gabès are given in **Table 1**, confirming the widespread status of eutrophication in the coastal waters of the Gulf of Gabès.

4. Signs of eutrophication in the Gulf of Gabès

Eutrophication yields a range of consequences impacting water quality, ecosystems, human health and economic activities [39].

Changes in coastal ecosystem states due to coastal eutrophication include:

- Presence of low dissolved oxygen and formation of hypoxic or "dead" zones (oxygen-depleted waters) particularly on the bottom [6, 40]. Hypoxia and anoxia significantly affect living resources and can cause severe damage to fisheries [41].

- A shift in the composition of phytoplankton species toward more tolerant and opportunistic species [42], creating favorable conditions for the proliferation of harmful algal blooms [43].

- A decrease in zooplankton diversity [44] as larger-sized taxa are replaced by smaller ones [45, 46].

- Loss of marine biodiversity of the aquatic community [47, 48] and the dominance of gelatinous organisms [49, 50].

- A reduction in subaquatic vegetation due to excessive macroalgae and microalgae growth reduces light penetration [40, 51].

The Gulf of Gabès has exhibited these characteristic signs of eutrophication exacerbated by trace metals pollution over the last decades.

4.1 Degradation of water quality in the Gulf of Gabès

The seawater temperature range from 13°C during winter to 26°C in the summer [52]. The mean annual salinity is notably high, around 38 psu, and may surpass 39 psu during summer [53]. The lowest recorded salinity (36 psu) and pH (7.6) were observed in the spring in areas adjacent to discharges from TCG and urban wastewaters [17].

Various studies have highlighted deteriorations in water quality, including turbidity [18, 30, 54] and depletion in seawater oxygen levels [30, 55].

4.2 Decline of *Posidonia oceanica* in the Gulf of Gabès

The characterization and evaluation of specific responses to eutrophication in seagrasses offer a valuable tool for detecting changes in water quality in coastal areas, especially considering they are among the most widespread organisms in coastal waters [56, 57]. Eutrophication has been cited as a primary factor leading to the global decline of seagrass populations [58]. Therefore, *Posidonia oceanica* is suggested for use as a bioindicator to assess the health status of coastal habitats [59, 60]. The surface alkaline phosphatase activity in the seagrass *P. oceanica* can be used as a biomarker of eutrophication [57].

At the beginning of the twentieth century, the littoral beds of the endemic Mediterranean seagrass *P. oceanica* almost entirely covered the sea floor in Gulf of Gabès [61]. However, the present condition of *P. oceanica* indicates an ongoing decline in its meadows and its localized disappearance from multiple areas within the central part of Gabès Gulf [62]. Consequently, a noticeable decline in associated fish production has been consistently observed since 1990 [63]. The loss of the native vegetation cover in the Gulf of Gabès is estimated at 90%, with the *P. oceanica* beds being replaced by the opportunistic green algae *Caulerpa prolifera* in deeper zones [64]. This decline is a global phenomenon and is supposed to be primarily induced by eutrophication [65]. In addition to the effects of eutrophication, the warming of the sea may lead to synergistic effects and an increased rate of loss for these valuable ecosystems [66, 67]. The decline of *P. oceanica* in the Gulf of Gabès is linked to pollution from the phosphate industries [33, 62].

4.3 Phytoplankton and harmful microalgae blooms in the Gulf of Gabès

In the coastal region of the Gulf of Gabès, the phytoplankton community is composed of seven major groups: Dinophyceae (108 taxa), Bacillariophyceae (58 taxa), Cyanobacteria (5 taxa), Dictyochophyceae (1 taxon), Euglenophyceae (1 taxon), Coccolithophorideae (1 taxon) and Chlorophyceae (1 taxon) [68]. Within the identified species, ten have been recognized as potentially toxic, including *Alexandrium minutum*, *Coolia monotis*, *Karenia selliformis*, *Protoceratium reticulatum* [68], *Amphidinium carterae*, *Dinophysis caudata*, *Prorocentrum lima*, *Prorocentrum minimum*, *Pseudonitzchia* sp. [17] and *Ostreopsis* cf. *ovata* [69].

Since 1990, multiple blooms of toxic dinoflagellates have been detected in the Gulf of Gabès [70–73]. Harmful algal blooms (HABs) of *A. minutum* formed in areas subjected to anthropogenic eutrophication such as Sfax Harbor and in confined lagoons in the Gulf [71]. The sudden *A. minutum* blooms along the nearshore of the Gulf of Gabès are complex, but phosphorus appears to be the key driving factor [71].

In 1994, *K. selliformis* was responsible for a significant intensive fish kill, estimated at 200 tons in the Gulf of Gabès [74, 75]. The occurrence of this taxon in the Gulf is related to high nitrate levels and is supported by elevated temperatures [76]. Typically, late summer and autumn are the periods of high bloom frequency of this taxon [70].

The abundance of Cyanobacteria in the Gulf of Gabès was positively correlated with inorganic nitrogen [70]. Blooms of *Trichodesmium erythraeum* were initially observed in July 1988 [77]. *T. erythraeum* blooms occurred approximately 2.11 times per year during the period from 1988 to 2013 [78]. Damages resulting from the toxic algal blooms cannot be considered resolved by reducing their populations. Indeed, numerous toxic Dinophyceae produce cysts at their blooming, successively accumulating in the sediments where they wait for the return of favorable conditions to germinate [79]. The encystment phenomenon is regular and frequent in the Gulf of Gabès but affects a low variety of species. During two decades of monitoring, eight different groups of cysts were recorded in the Gulf, with the dominance of ones of potentially toxic species [80, 81]. Even in the absence of algal inoculation from neighboring geographic areas, such toxic species can re-appear by the germination of the cysts in the sediments each time the conditions are favorable. Thus, the management of coastal areas has to carefully consider this "potential" harmful injection, even in the absence of algal blooms in the water column for many years.

4.4 Copepoda in the Gulf of Gabès

Zooplankton assemblages in the Gulf of Gabès were primarily dominated by copepods, accounting for 69–83% of the total zooplankton abundance [82]. A total of 52 species of Calanoida, 30 Cyclopoida and 11 Harpacticoida species were reported in the Gulf of Gabès [83, 84]. Notably, Cyclopoida, particularly Oithonidae (mainly *Oithona nana*), dominated copepod communities in the Gulf of Gabès [17, 35, 36, 82, 85]. We suggest that with increasing eutrophication over decades, a general shift to smaller-sized Copepoda with egg sacs, particularly the cyclopoid Oithona nana, was observed in the Gulf of Gabès. The success of *Oithona* spp. in eutrophicated and disturbed ecosystems is mainly due to their remarkable adaptability to changing environments than other species [86] as they are typical ecological generalists [83, 84]. In fact, the dominance of this species can be linked to a change in the food spectrum (and/or the fact that it does not spawn eggs into the water column). The possession of egg sacs is considered as an advantage in eutrophicated ecosystems that evolve hypoxic/anoxic bottom waters [7]. Spawned eggs that fall into the sediments may be prevented from hatching due to reduced oxygen levels. Therefore, the combination of a successful reproductive strategy, an omnivorous diet, and lower metabolic demands likely underlines the prominence of *Oithona* in the Gulf of Gabès [85, 87].

5. Conclusion

As documented by a number of investigations, eutrophication is a serious problem for the biodiversity in the Gulf of Gabès. However, since the Gulf also experienced chemical pollution by trace metals, and overfishing, eutrophication may not have been the only factor. The enrichment of nutrients, especially phosphorus, interacts with these factors and, in addition, with climate change. Due to cumulative anthropogenic pressures, integrated management of the coastal marine environment of the Gulf of Gabès remains necessary to re-direct the ecosystem functioning toward a healthier status and a restoration of ecosystem services [88, 89] possibly compromised. Efforts in the Gulf of Gabès to contain eutrophication involve a policy of environmentally sustainable development, with tighter industrial regulations, enhanced agriculture and improved wastewater management. The heavy socio-economic impact of such a policy is only apparent. In fact, the decision should be sustained by detailed studies on the economic convenience of such an approach, with a precise monetization of what the present situation subtracts in terms of ecosystem services to society. The restoration of good environmental status, in addition, would not be a cost for managers because open gulfs (such as the Gabès one) should be easily re-populated. Living organisms, in fact, can go elsewhere when local conditions are adverse or can undergo a lethargic condition, waiting for the return of suitable conditions in the sediments for years [90, 91]. The re-covering of a good environmental status (and the re-launch of ecosystem services) could be, consequently, a fast process.

Author details

Neila Annabi-Trabelsi[1]*, Mohammad Ali[2], Genuario Belmonte[3,4], Habib Ayadi[1] and Wassim Guermazi[1]

1 Marine Biodiversity and Environment Laboratory (LR18ES30), University of Sfax, Sfax CP, Tunisia

2 Environment and Life Sciences Research Center, Kuwait Institute for Scientific Research, Safat, Kuwait

3 Laboratory of Zoogeography and Fauna, Department of Biological and Environmental Sciences and Technologies, University of Salento, Lecce, Italy

4 National Biodiversity Future Center NBFC—CNR, Palermo, Italy

*Address all correspondence to: neila.trabelsi@isbs.usf.tn

References

[1] Smith VH. Using primary productivity as an index of coastal eutrophication: The units of measurement matter. Journal of Plankton Research. 2006;**29**(1):1-6

[2] Chen X, Wang Y, Cai Z, Zhang M, Ye C. Response of the nitrogen load and its driving forces in estuarine water to dam construction in Taihu Lake, China. Environmental Science and Pollution Research International. 2020;**27**(25):31458-31467

[3] Wurtsbaugh WA, Paerl HW, Dodds WK. Nutrients, eutrophication and harmful algal blooms along the freshwater to marine continuum. WIREs Water. 2019;**6**(5):e1373

[4] Directive EUW. Council directive of 21 May 1991 concerning urban waste water treatment (91/271/EEC). Journal of European Communities. 1991;**34**:40

[5] Schindler DW. Eutrophication and recovery in experimental lakes: Implications for lake management. Science. 1974;**184**(4139):897-899

[6] Malone TC, Newton A. The globalization of cultural eutrophication in the coastal ocean: Causes and consequences. Frontiers in Marine Science. 2020;7:670

[7] Marcus N. An overview of the impacts of eutrophication and chemical pollutants on copepods of the coastal zone. Zoological Studies. 2004;**43**(2):211-217

[8] Glibert PM, Burford MA. Globally changing nutrient loads and harmful algal blooms: Recent advances, new paradigms, and continuing challenges. Oceanography. 2017;**30**(1):58-69

[9] Primpas I, Karydis M. Scaling the trophic index (TRIX) in oligotrophic marine environments. Environmental Monitoring and Assessment. 2011;**178**(1-4):257-269

[10] Karydis M, Kitsiou D. Eutrophication and environmental policy in the Mediterranean Sea: A review. Environmental Monitoring and Assessment. 2012;**184**:4931-4984

[11] Group TM, de Madron XD, Guieu C, Sempéré R, Conan P, Cossa D, et al. Marine ecosystems' responses to climatic and anthropogenic forcings in the Mediterranean. Progress in Oceanography. 2011;**91**(2):97-166

[12] Unep/Map. State of the Mediterranean marine and coastal environment. In: United Nations Environment Programme/Mediterranean Action Plan (UNEP/MAP). Athens: Barcelona Convention; 2012. p. 96

[13] Yilmaz A, Baesturk O, Ediger D, Ylmaz K, Hatipoglu E. Eutrophication in Iskenderun Bay, North-Eastern Mediterranean. In: Science of the Total Environment. Amsterdam: Elsevier; 1992. pp. 705-717

[14] Moncheva S, Gotsis-Skretas O, Pagou K, Krastev A. Phytoplankton blooms in Black Sea and Mediterranean coastal ecosystems subjected to anthropogenic eutrophication: Similarities and differences. Estuarine, Coastal and Shelf Science. 2001;**53**(3):281-295

[15] Pavlidou A, Simboura N, Rousselaki E, Tsapakis M, Pagou K, Drakopoulou P, et al. Methods of eutrophication assessment in the

context of the water framework directive: Examples from the eastern Mediterranean coastal areas. Continental Shelf Research. 2015;**108**:156-168

[16] Viaroli P, Nizzoli D, Pinardi M, Soana E, Bartoli M. Eutrophication of the Mediterranean Sea: A watershed—Cascading aquatic filter approach. Rendiconti Lincei. Scienze Fisiche e Naturali. 2015;**26**(1):13-23

[17] Annabi-Trabelsi N, Guermazi W, Leignel V, Al-Enezi Y, Karam Q, Ali M, et al. Effects of eutrophication on plankton abundance and composition in the Gulf of Gabès (Mediterranean Sea, Tunisia). Water. 2022;**14**(14):2230

[18] Katlane R, Nechad B, Ruddick K, Zargouni F. Optical remote sensing of turbidity and total suspended matter in the Gulf of Gabes. Arabian Journal of Geosciences. 2013;**6**(5):1527-1535

[19] El Zrelli R, Rabaoui L, Alaya MB, Daghbouj N, Castet S, Besson P, et al. Seawater quality assessment and identification of pollution sources along the central coastal area of Gabes gulf (SE Tunisia): Evidence of industrial impact and implications for marine environment protection. Marine Pollution Bulletin. 2018;**127**:445-452

[20] Rabaoui L, Balti R, Zrelli R, Tlig-Zouari S. Assessment of heavy metals pollution in the gulf of Gabes (Tunisia) using four mollusk species. Mediterranean Marine Science. 2013;**15**(1):45

[21] El Zrelli R, Courjault-Radé P, Rabaoui L, Castet S, Michel S, Bejaoui N. Heavy metal contamination and ecological risk assessment in the surface sediments of the coastal area surrounding the industrial complex of Gabes city, Gulf of Gabes, SE Tunisia. Marine Pollution Bulletin. 2015;**101**(2):922-929

[22] Sammari C, Koutitonsky VG, Moussa M. Sea level variability and tidal resonance in the Gulf of Gabes, Tunisia. Continental Shelf Research. 2006;**26**(3):338-350

[23] Othmani A, Béjaoui B, Chevalier C, Elhmaidi D, Devenon JL, Aleya L. High-resolution numerical modelling of the barotropic tides in the Gulf of Gabes, eastern Mediterranean Sea (Tunisia). Journal of African Earth Sciences. 2017;**129**:224-232

[24] Mezghani-Chaari S, Hamza A, Hamza-Chaffai A. Mercury contamination in human hair and some marine species from Sfax coasts of Tunisia: Levels and risk assessment. Environmental Monitoring and Assessment. 2011;**180**:477-487

[25] Gargouri D, Gzam M, Kharroubi A, Jedoui Y. Use of sediment quality indicators for heavy metals contamination and ecological risk assessment in urbanized coastal zones. Environmental Earth Sciences. 2018;**77**:1-8

[26] Kharroubi A, Gzam M, Jedoui Y. Anthropogenic and natural effects on the water and sediments qualities of costal lagoons: Case of the Boughrara Lagoon (Southeast Tunisia). Environmental Earth Sciences. 2012;**67**:1061-1067

[27] Bejaoui B, Rais S, Koutitonsky V. Modélisation de la dispersion du phosphogypse dans le golfe de Gabès. INSTM Bulletin: Marine and Freshwater Sciences. 2004;**31**:113-119

[28] Rutherford PM, Dudas MJ, Samek RA. Environmental impacts of phosphogypsum. Science of the Total Environment. 1994;**149**(1-2):1-38

[29] Ajmal PY, Bhangare RC, Tiwari M, Sahu SK, Pandit GG. External gamma

radiation levels and natural radioactivity in soil around a phosphate fertilizer plant at Mumbai. Journal of Radioanalytical and Nuclear Chemistry. 2014;**300**:23-27

[30] El Kateb A, Stalder C, Rüggeberg A, Neururer C, Spangenberg JE, Spezzaferri S. Impact of industrial phosphate waste discharge on the marine environment in the Gulf of Gabes (Tunisia). PLoS One. 2018;**13**(5):e0197731

[31] Correll DL. The role of phosphorus in the eutrophication of receiving waters: A review. Journal of Environmental Quality. 1998;**27**(2):261-266

[32] Ngatia L, Grace JM III, Moriasi D, Taylor R. Nitrogen and phosphorus eutrophication in marine ecosystems. Monitoring of Marine Pollution. 2019;**1**:1-17

[33] Brahim MB, Hamza A, Hannachi I, Rebai A, Jarboui O, Bouain A, et al. Variability in the structure of epiphytic assemblages of Posidonia oceanica in relation to human interferences in the Gulf of Gabes. Tunisia. Marine Environmental Research. 2010;**70**(5):411-421

[34] Ignatiades L, Karydis M, Vounatsou P. A possible method for evaluating oligotrophy and eutrophication based on nutrient concentration scales. Marine Pollution Bulletin. 1992;**24**(5):238-243

[35] Salem ZB, Drira Z, Ayadi H. What factors drive the variations of phytoplankton, ciliate and mesozooplankton communities in the polluted southern coast of Sfax, Tunisia? Environmental Science and Pollution Research. 2015;**22**:11764-11780

[36] Drira Z, Kmiha-Megdiche S, Sahnoun H, Tedetti M, Pagano M, Ayadi H. Copepod assemblages as a bioindicator of environmental quality in three coastal areas under contrasted anthropogenic inputs (Gulf of Gabes, Tunisia). Journal of the Marine Biological Association of the United Kingdom. 2018;**98**(8):1889-1905

[37] Primpas I, Tsirtsis G, Karydis M, Kokkoris GD. Principal component analysis: Development of a multivariate index for assessing eutrophication according to the European water framework directive. Ecological Indicators. 2010;**10**(2):178-183

[38] Sahraoui I, Grami B, Bates SS, Bouchouicha D, Chikhaoui MA, Mabrouk HH, et al. Response of potentially toxic pseudo-nitzschia (Bacillariophyceae) populations and domoic acid to environmental conditions in a eutrophied, SW Mediterranean coastal lagoon (Tunisia). Estuarine, Coastal and Shelf Science. 2012;**102**:95-104

[39] Hwang SJ. Eutrophication and the ecological health risk. International Journal of Environmental Research and Public Health IJERPH. 2020;**17**(17):6332

[40] Rabalais NN, Cai WJ, Carstensen J, Conley DJ, Fry B, Hu X, et al. Eutrophication-driven deoxygenation in the coastal ocean. Oceanography. 2014;**27**(1):172-183

[41] Karim MR, Sekine M, Ukita M. Simulation of eutrophication and associated occurrence of hypoxic and anoxic condition in a coastal bay in Japan. Marine Pollution Bulletin. 2002;**45**(1-12):280-285

[42] McQuatters-Gollop A, Gilbert AJ, Mee LD, Vermaat JE, Artioli Y, Humborg C, et al. How well do ecosystem indicators communicate the effects of anthropogenic eutrophication? Estuarine, Coastal and Shelf Science. 2009;**82**(4):583-596

[43] Brush MJ, Giani M, Totti C, Testa JM, Faganeli J, Ogrinc N, et al. Eutrophication, harmful algae, oxygen depletion, and acidification. In: Coastal Ecosystems in Transition: A Comparative Analysis of the Northern Adriatic and Chesapeake Bay. In: Malone TC, Malej A, Faganeli J, editors. Geophysical Monograph Series 256. American Geophysical Union. 1st ed. Hoboken, NJ, USA: John Wiley and Sons; 2020. pp. 75-104

[44] Loanna S-F, Papathanassiou E, Lepretre A, Frontier S. Zooplankton assemblages and influence of environmental parameters on them in a Mediterranean coastal area. Journal of Plankton Research. 1998;**20**(5):847-870

[45] Uye S-I. Replacement of large copepods by small ones with eutrophication of embayments: Cause and consequence. In: Hydrobiologia. Hamburg: Springer; 1994;**292/293**:513-519

[46] Belmonte G, Moscatello S, Hajdëri E, Vaglio I, Denitto F. Composition and spatial distribution of mesozooplankton along confinement and anthropogenic-impact gradients in the gulf of Vlore (Albania). Journal of Coastal Research. 2018;**34**(1):174-184

[47] Horta PA, Rörig LR, Costa GB, Baruffi JB, Bastos E, Rocha LS, et al. Marine eutrophication: Overview from now to the future. In: Häder DP, Helbling EW, Villafañe VE, editors. Anthropogenic Pollution of Aquatic Ecosystems. Cham: Springer; 2021. pp. 157-180. DOI: 10.1007/978-3-030-75602-4_8

[48] Fetahi T. Eutrophication of Ethiopian water bodies: A serious threat to water quality, biodiversity and public health. African Journal of Aquatic Science. 2019;**44**(4):303-312

[49] Richardson AJ, Bakun A, Hays GC, Gibbons MJ. The jellyfish joyride: Causes, consequences and management responses to a more gelatinous future. Trends in Ecology & Evolution. 2009;**24**(6):312-322

[50] Hashim AR, Kamaruddin SA, Buyong F, Mat Nazir EN, Che Ismail CZ, Tajam J, et al. Jellyfish blooming: Are we responsible? In: ICAN International Virtual Conference 2022 (IIVC 2022) Proceedings–Navigating the VUCA World: Harnessing the Role of Industry Linkages, Community Development and Alumni Network in Academia. Arau, Malaysia; 2022. pp. 51-60

[51] Adams JB, Taljaard S, Van Niekerk L, Lemley DA. Nutrient enrichment as a threat to the ecological resilience and health of south African microtidal estuaries. African Journal of Aquatic Science. 2020;**45**(1-2):23-40

[52] Béjaoui B, Ben Ismail S, Othmani A, Hamida OBA-BH, Chevalier C, Feki-Sahnoun W, et al. Synthesis review of the Gulf of Gabes (eastern Mediterranean Sea, Tunisia): Morphological, climatic, physical oceanographic, biogeochemical and fisheries features. Estuarine, Coastal and Shelf Science. 2019;**219**:395-408

[53] Mounir BB, Asma H, Sana BI, Lotfi M, Abderrahmen B, Lotfi A. What factors drive seasonal variation of phytoplankton, protozoans and metazoans on leaves of Posidonia oceanica and in the water column along the coast of the Kerkennah Islands, Tunisia? Marine Pollution Bulletin. Amsterdam: Elsevier; 2013;**71**(1-2):286-298

[54] Darmoul B. Pollution Dans le Golfe de Gabès (Tunisie): Bilan de Six Années de Surveillance (1976-1981). Institut National des Sciences et Technologies de la Mer; 1988;**15**:61-84

[55] Mosbahi N, Pezy JP, Dauvin JC, Neifar L. COVID-19 pandemic lockdown: An excellent opportunity to study the effects of trawling disturbance on macrobenthic Fauna in the shallow waters of the Gulf of Gabès (Tunisia, Central Mediterranean Sea). International Journal of Environmental Research and Public Health. 2022;**19**(3):1282

[56] Spalding M, Taylor M, Ravilious C, Short FT, Green E. Global overview: The distribution and status of seagrasses. In: Green EP, Short FT, editors. World Atlas of Seagrasses. 1st edition. Berkeley: University of California Press; 2003. pp. 5-26

[57] Martínez-Crego B, Romero J, Alcoverro T. The use of surface alkaline phosphatase activity in the seagrass Posidonia oceanica as a biomarker of eutrophication. Marine Ecology. 2006;**27**(4):381-387

[58] Green EP, Short FT. World Atlas of Seagrasses. Prepared by the UNEP World Conservation Monitoring Centre. 1st ed. Berkeley, USA: University of California Press; 2003. pp. 1-324

[59] Cannac M, Ferrat L, Pergent-Martini C, Pergent G, Pasqualini V. Effects of fish farming on flavonoids in Posidonia oceanica. Science of the Total Environment. 2006;**370**(1):91-98

[60] Bonanno G, Di Martino V. Trace element compartmentation in the seagrass Posidonia oceanica and biomonitoring applications. Marine Pollution Bulletin. 2017;**116**(1-2):196-203

[61] Zaouali J. Les peuplements benthiques de la petite Syrte, golfe de Gabès-Tunisie. Résultats de la campagne de prospection du mois de juillet 1990, Etude préliminaire: Biocénoses et thanatocénoses récentes. Life. 1990;**3**(1-2):47-60

[62] El Zrelli R, Courjault-Radé P, Rabaoui L, Daghbouj N, Mansour L, Balti R, et al. Biomonitoring of coastal pollution in the Gulf of Gabes (SE, Tunisia): Use of Posidonia oceanica seagrass as a bioindicator and its mat as an archive of coastal metallic contamination. Environmental Science and Pollution Research. 2017;**24**:22214-22225

[63] Hamza-Chaffai A, Pellerin J, Amiard JC. Health assessment of a marine bivalve Ruditapes decussatus from the Gulf of Gabès (Tunisia). Environment International. 2003;**28**(7):609-617

[64] Ramos-Espla A, Ouergui A, Bayle JT, Ben Mbarek N, Fernanadez-Troquemada Y, Gallouza S, et al. Contribution à la caracterisation des herbiers à Posidonia oceanica (L.) Delile aux iles Kerkennah (est de la Tunisie, mediterranée orientale). In: Proceedings of the First Mediterranean Symposium on Marine Vegetation. Ajaccio, France; 2000. pp. 177-181

[65] Touchette BW, Burkholder JM, Glasgow HB. Variations in eelgrass (Zostera marina L.) morphology and internal nutrient composition as influenced by increased temperature and water column nitrate. Estuaries. 2003;**26**:142-155

[66] Lloret J, Marín A, Marín-Guirao L. Is coastal lagoon eutrophication likely to be aggravated by global climate change? Estuarine, Coastal and Shelf Science. 2008;**78**(2):403-412

[67] Pazzaglia J, Santillán-Sarmiento A, Helber SB, Ruocco M, Terlizzi A, Marín-Guirao L, et al. Does warming enhance the effects of eutrophication in the seagrass Posidonia oceanica? Frontiers in Marine Science. 2020;**7**:564805

[68] Feki-Sahnoun W, Hamza A, Béjaoui B, Mahfoudi M, Rebai A, Bel HM. Multi-table approach to assess the biogeography of phytoplankton: Ecological and management implications. Hydrobiologia. 2018;**815**:229-251

[69] Abdennadher M, Zouari AB, Sahnoun WF, Alverca E, Penna A, Hamza A. Ostreopsis cf. ovata in the Gulf of Gabès (South-Eastern Mediterranean Sea): Morphological, molecular and ecological characterization. Harmful Algae. 2017;**63**:56-67

[70] Drira Z, Hamza A, Belhassen M, Ayadi H, Bouaïn A, Aleya L. Dynamics of dinoflagellates and environmental factors during the summer in the Gulf of Gabes (Tunisia, eastern Mediterranean Sea). Scientia Marina. 2008;**72**(1):59-71

[71] Abdenadher M, Hamza A, Fekih W, Hannachi I, Bellaaj AZ, Bradai MN, et al. Factors determining the dynamics of toxic blooms of Alexandrium minutum during a 10-year study along the shallow southwestern Mediterranean coasts. Estuarine, Coastal and Shelf Science. 2012;**106**:102-111

[72] Walha LD, Hamza A, Keskes FA, Cibic T, Mechi A, Mahfoudi M, et al. Heavy metals accumulation in environmental matrices and their influence on potentially harmful dinoflagellates development in the Gulf of Gabes (Tunisia). Estuarine, Coastal and Shelf Science. 2021;**254**:107317

[73] Feki-Sahnoun W, Njah H, Abdennadher M, Hamza A, Barraj N, Mahfoudi M, et al. A Naïve Bayesian network approach to determine the potential drivers of the toxic dinoflagellate Coolia monotis (Meunier, 1919) in the Gulf of Gabès, Tunisia. Euro-Mediterranean Journal for Environmental Integration. 2019;**4**:1-11

[74] Hamza A, El Abed A. Les eaux colorées dans le golfe de Gabes. Bulletin de l'Institut National Scientifique et Technique d'Océanographie et de Pêche de Salammbo. 1994;**21**:66-77

[75] Souâd T, El Abed A. On the presence of potentially toxic algae in the lagoons of Tunisia. Harmful Algae News. 2001;**22**:10

[76] Feki W, Hamza A, Frossard V, Abdennadher M, Hannachi I, Jacquot M, et al. What are the potential drivers of blooms of the toxic dinoflagellate Karenia selliformis? A 10-year study in the Gulf of Gabes, Tunisia, southwestern Mediterranean Sea. Harmful Algae. 2013;**23**:8-18

[77] Hamza A, Naceur BEN. Sur l'apparition du phénomène" d'eau rouge" dans le golfe de Gabès en été 1988. INSTM Bulletin: Marine and Freshwater Sciences. 1990;**17**:5-16

[78] Sabeur HI, Wafa FS, Asma H, Malika BH. Long term characterization of Trichodesmium erythraeum blooms in Gabès gulf (Tunisia). Continental Shelf Research. 2016;**124**:95-103

[79] Belmonte G, Rubino F. Cysts and resting eggs from marine zooplankton: Dimension of the phenomenon, physiology of rest, and ecological and biogeographic implications. In: Alekseev VR, Pinel-Alloul B, editors. Dormancy in Aquatic Organisms Theory, Human Use and Modeling [Internet], Monographiae Biologicae. Vol. 92. Cham: Springer International Publishing; 2019. pp. 71-94. Available from: http://link.springer.com/10.1007/978-3-030-21213-1_5

[80] Keskes FA, Ayadi N, Atoui A, Mahfoudi M, Abdennadher M, Walha LD, et al. Dinoflagellates encystment with emphasis on blooms in Boughrara lagoon (South-Western

Mediterranean): Combined effects of trace metal concentration and environmental context. Estuarine, Coastal and Shelf Science. 2020;**237**:106648

[81] Abdelmouleh F, Hamza A, Mahfoudi M, Walha LD, Barraj N, Hassen MB. Encystment of dinoflagellate in tidal coasts of Gabes gulf (Tunisia): Analyse of long term observations. INSTM Bulletin: Marine and Freshwater Sciences. 2018;**45**:23-26

[82] Drira Z, Belhassen M, Ayadi H, Hamza A, Zarrad R, Bouaïn A, et al. Copepod community structure related to environmental factors from a summer cruise in the Gulf of Gabès (Tunisia, eastern Mediterranean Sea). Journal of the Marine Biological Association of the United Kingdom. 2010;**90**(1):145-157

[83] Annabi-Trabelsi N, Guermazi W, Ayadi H. Cyclopoida and Harpacticoida (crustacea: Copepoda) of the Gulf of Gabès: A review. Thalassia Salentina. 2020;**2020**(42):93-98

[84] Annabi-Trabelsi N, Guermazi W, Ayadi H. Marine Calanoida (crustacea: Copepoda) of the Gulf of Gabes: A review. Thalassia Salentina. 2020;**2020**(42):99-106

[85] Drira Z, Bel Hassen M, Ayadi H, Aleya L. What factors drive copepod community distribution in the Gulf of Gabes, eastern Mediterranean Sea? Environmental Science and Pollution Research. 2014;**21**(4):2918-2934

[86] Chihoub S, Christaki U, Chelgham S, Amara R, Ramdane Z, Zebboudj A, et al. Coastal eutrophication as a potential driver of functional homogenization of copepod species assemblages in the Mediterranean Sea. Ecological Indicators. 2020;**115**:106388

[87] Ben Ltaief T, Drira Z, Hannachi I, Bel Hassen M, Hamza A, Pagano M, et al. What are the factors leading to the success of small planktonic copepods in the Gulf of Gabes, Tunisia? Journal of the Marine Biological Association of the United Kingdom. 2015;**95**(4):747-761

[88] Pepper IL, Gerba CP, Brusseau ML. Environmental and Pollution Science. Amsterdam: Elsevier; 2011

[89] Costanza R, d'Arge R, De Groot R, Farber S, Grasso M, Hannon B, et al. The value of the world's ecosystem services and natural capital. Nature. 1997;**387**(6630):253-260

[90] Kerfoot WC, Weider LJ. Experimental paleoecology (resurrection ecology): Chasing Van Valen's red queen hypothesis. Limnology and Oceanography. 2004;**49**(4 part 2):1300-1316

[91] Belmonte G, Rubino F. Resting cysts from coastal marine plankton. Oceanography and Marine Biology. An Annual Review. In: Hawkins SJ, Allcock AL, Bates AE, Firth LB, Smith IP, Swearer SE, et al, editors. Boca Raton: CRC Press; Vol. 57. 2019. pp. 1-88

Chapter 2

Biostimulant Properties of Marine Bioactive Extracts in Plants: Incrimination toward Sustainable Crop Production in Rice

Melekote Nagabhushan Arun, Rapolu Mahender Kumar, Sailaja Nori, Banugu Sreedevi, Guntupalli Padmavathi, Pallakonda Revathi, Neha Pathak, Dayyala Srinivas and Raman Meenakshi Sundaram

Abstract

Enhancing productivity through integrated and comprehensive nutrient management is pertinent to sustainable intensification of agricultural ecosystems. The utilization of marine bioactive stimulants has been gaining momentum and impetus in crop agricultural farming system due to their phytoelicitor activity. Liquids biostimulants derived from seaweed evoke defense responses in plants that contribute to resistance to abiotic stresses and challenges like high temperature, salinity, moisture stress, and cold. Seaweed extracts are immensely organic and suitable for growing crops that are both organic and environmentally friendly. Seaweeds provide an abundant source of natural growth substances that can be employed to enhance plant growth. Seaweeds are one of the most significant marine resources of the world, and derived compounds have been extensively used as amendments in crop production systems due to the presence of macronutrients such as Ca, K, and P and micronutrients like Fe, Cu, Zn, B, Mn, Co, and Mo, presence of several plant growth stimulating compounds including cytokinin, auxins, gibberellins, and betaines which are essential for plant growth and development. The purpose of the current chapter is to explore the functional and growth characteristics induced by seaweed extracts in addition to their modes and mechanisms of action in rice crops, which are responsible for elicitor and phytostimulatory activities and boost in grain production and nutrient usage efficiency.

Keywords: seaweed extract, macroalgae, liquid biostimulants, stress tolerance, sustainable agriculture, phytoelicitor compounds

1. Introduction

Abiotic and biotic stresses restrict the growth and productivity of crops. The global effects of negative climatic changes have manifested as desertification, escalate atmospheric CO_2, temperature, and soil salanization, and nutrient imbalances like mineral toxicity and deficiency have caused dramatic effects on agricultural production and the quality of crops. Abiotic stresses have reduced the growth, development, productivity, and quality of plants and local extinction of species. The global amount of cultivable land for agriculture is continuously shrinking due to adverse effects of climate change. Abiotic stresses may be prevented by optimizing plant growth conditions and through provision of water, plant growth regulators (PGRs – auxins, cytokinins, gibberellins, stringolactones, and brassinosteroids) and nutrients. Rice is one of the most significant food crops grown across the world. It is a primary staple food for more than 50% of the world's population, accounting for more than 20% of their daily caloric consumption from rice [1]. Rice is cultivated in 160 million ha area globally with an average yield of 4631 kg ha^{-1} [2].

Biostimulants are increasingly being incorporated into production system with the objective of altering physiological processes in plants to optimize plant productivity [3]. Biostimulants present a potentially novel approach for regulating and altering physiological processes in plants to stimulate growth, to mitigate stress limitations, and to boost yield. Biostimulant materials are natural products which contain no added chemicals or synthetic plant growth regulators and which have a beneficial effect on plant growth. Seaweed/sea plants are gaining considerable importance due to the distribution, renewable nature, and widespread application [4]. Organic biostimulants are compounds which increase plant growth and vigor through increased efficiency of nutrient and water uptake. Plant biostimulants are compounds when applied to plants, seeds, or growing substrates have the capacity to modify physiological processes of plants in a way that provides potential advantages to growth and stress response. Seaweed (macroalgae) is a multicellular macroscopic algae species which are subdivided into red, green, and brown algae due to their pigmentation according to phyla, Rhodophyta, Chlorophyta, and Phaeophyceae. Nine thousand species of macroalgae are broadly classified into three main groups, viz. Rhodophyta (red algae), Ochrophyta-Phaeophyceae (Brown algae), and Chlorophyta (green algae), based on pigments present. Sea plant biostimulants derived from natural raw resources are used in the ultrasmall and minute doses for modification of biochemical and physiological plant processes with the objective of more complete realization of genetic potential of their productivity due to changes in hormonal status, activation of metabolic processes, increase in nutrient efficiency, stimulation of growth, development, and strengthening of the potential to withstand the negative effects of various stress factors. Seaweed biostimulants are new class of inputs offering a potential alternative to traditional organic chemical inputs and can reduce the application of fertilizers to improve soil health and increase soil microbial activity [5]. Chemical analysis of seaweeds and their extracts have revealed the presence of a broad spectrum of plant growth regulators/phyto-hormones such as auxins, gibberellins, cytokinins, and betaines, enzymes, vitamins and hydrolyzed proteins, polysaccharides, nutrients, and trace elements (Fe, Cu, Zn, Co, Mo, Mn, and Ni) in varying amounts and biologically active alginic acids, polyphenols, and free amino acids which invigorate plant growth and alleviating several biotic and abiotic stresses [6]. Seaweeds often contain larger concentrations of Ca, K, Mg, Na, Cu, Fe, and Zn than terrestrial plants [5]. The biostimulant aids in fostering the development of beneficial soil microorganisms, developing tolerance to environmental stress and improving nutrient uptake from soil and enhancing antioxidant properties. Seaweed extract supplement improved significantly the content of nitrate nitrogen (N), available phosphorus (P), and available potassium (K) in rhizosphere soil in the tillering stage, and, finally, increased the rice yield and quality mildly in rice crop.

Biostimulants are agents which at very low concentrations improve the basic biochemical processes in plants and soil and thereby improve the growth and development of plants and increase resistance to stress. The seaweed properties with most potential relevance to crop production involve elemental composition (e.g. primary plant nutrients, other nutrient elements, heavy metals, and salts) and organic compound composition (e.g. energy sources for microbial processes). Seaweed extracts improved the number and size of chloroplasts, the growth of chloroplast grana, and the concentration of chlorophyll in leaves due to the presence of beneficial amounts of cytokinins, auxins, betaines, and inorganic salts [7]. The most commonly used seaweed as a biostimulant is an alga, obtained from North Atlantic Ocean and known as *Ascophyllum nodosum*, rich in polysaccharides (alginate, fucoidan, and laminarin), minerals, and vitamins [8]. The goal of the chapter is to provide comprehensive analysis of the current situation in the field of biostimulants and to develop a science-based theoretical foundation for the conceptualization, classification, and practical application of sea plant biostimulants. The structure of the chapter is based on the consideration of biostimulants in terms of the action on different regulatory and functional system of plants (signaling, metabolism, uptake, and transport system) using both conceptual and methodological approaches.

2. Different types of seaweed/plants and its distribution

The seaweeds are distributed horizontally in different zonations, viz. supra-tidal (supra-littoral), intertidal (littoral) and subtidal (sub-littoral) regions of the seas and oceans. Green seaweeds are most commonly found in the intertidal zone (**Table 1**). Common green seaweeds are species of *Ulva* (sea lettuce), *Enteromorpha* (green string lettuce), *Chaetomorpha*, *Codium*, and *Caulerpa*. Brown seaweeds inhabit in the tidal or upper subtidal zone. Common brown seaweeds are species of *Sargassum*, *Laminaria*, *Turbinaria*, and *Dictyota*. Red seaweeds grow in subtidal waters. Common red seaweeds are species of *Gracilaria*, *Gelidiella*, *Eucheuma*, *Ceramium*, and *Acanthophora*. The blue green algae grow in supra-tidal region mostly as colonies, and sometimes they occur as epiphytes on other algae. Common blue green algae are species of *Lyngbya*, *Spirulina*, and *Oscillatora*. Seaweeds, which are multicellular, microscopic plants that are found coastal locations, are an integral component of coastal marine ecosystems. Marine algae or seaweeds are classified into three taxonomic groups that have diverse pigment composition: Rhodophyta (red algae), Chlorophyta (green algae), and Ochrophyta (brown algae). The Indian Ocean is host to 865 species of seaweed belonging to 216 genera and 68 families [9]. Brown algae (Phaeophyceae) are the largest prevalent type of seaweed that are habitat to water of temperate region where more than 2000 species are found. Green algae (Chlorophytta) can be found in marine or freshwater habitats, and some even thrive in moist soils where more than 1800 species are found. Green algae come in three forms: unicellular, colonial, or multicellular. Sea lettuce (*Ulva lactuca*) is a type of green algae commonly found in tidal pools. Red seaweed (Rodophyta) is the largest group of algae in the plant kingdom which account more than 7200 species

Class	Order	Family	*Species*
Rhodophyceae (Red algae)	Ceramiales	Ceramiaceae	*Spyridia filamentosa Ceramium planum*
		Rhodomelaceae	*Amansia glomerata Laurencia papillosa Acanthophora spicifera Laurencia johnstonii Laurencia obtusa*
	Corallinales	Lithophyllaceae	*Amphiroa rigida Amphiroa gracilis*
		Mastoporaceae	*Mastophora rosea Mastophora pacifica Mastophora multistrata*
		Corallinaceae	*Corallina elongate Janie rubens Lithothamnium calcareum*
	Nemaliales	Galaxauraceae	*Galaxaura apiculata Galaxaura fastigiata Galaxaura filamentosa Actinotrichia fragilis*
		Acrochaetaceae	*Porphyra perforate*
	Cyanidiales	Cyanidiaceae	*Cyanidium caldarium*
	Gelidiales	Gelidiaceae	*Gelidium serrulatum*
	Gracilariales	Gracilariaceae	*Gracilaria edulis Gracilaria gracilis Gracilaria salicornia Gracilaria textorii Gracilaria verrucos Gracilaria dura*
	Gigartinales	Solieriaceae	*Eucheuma denticulatum Eucheuma spinosum Kappaphycus alvarezii Kappaphycus cottonii Sarcodiotheca chordalis*
		Cystocloniaceae	*Hypnea spinella Hypnea musciformis*
Phaeophyceae (Brown algae)	Dictyotales	Dictyotaceae	*Dictyota dichotoma Padina australis Padina pavonica*
	Ectocarpales	Scytosiphonaceae	*Hydroclathrus clathratus Hydroclathrus minutus*
	Fucales	Sargassaceae	*Sargassum cristaefolium Sargassum polycystum Cystoseira humilis*
		Fucaceae	*Ascophyllum nodosum Fucus gardneri Fucus spiralis Fucus vesiculosus*
		Alariaceae	*Alaria esculenta*
		Durvillaeaceae	*Durvillea Antarctica Durvillea protatorum*
	Laminariales	Lessoniaceae	*Ecklonia maxima Nereocystis luetkeana*
		Laminariaceae	*Laminaria digitata Macrocycstis pyrifera*

Class	Order	Family	*Species*
Ulvophyceae (Green algae)	Bryopsidales	Caulerpaceae	*Caulerpa paspaloides Caulerpa lentillifera Caulerpa racemosa Caulerpa serrulata Caulerpa sertularioides*
		Halimedaceae	*Halimeda cuneata Halimeda cylindracea Halimeda discoidea Halimeda incrassata Halimeda macroloba Halimeda opuntia Halimeda tuna*
		Udoteaceae	*Udotea geppiorum*
		Dichotomosiphonaceae	*Avrainvillea erecta Avrainvillea lacerata*
		Codiaceae	*Codium ovale Codium bursa Codium liyengarii Codium tomentosum*
	Cladophorales	Cladophoraceae	*Chaetomorpha crassa Chaetomorpha spiralis Cladophora rupestris*
		Anadyomenaceae	*Anadyomene wrightii Microdictyon marinum*
		Siphonocladaceae	*Boergesenia forbesii Dictyosphaeria cavernosa*
		Valoniaceae	*Valonia aegagropila Valonia fastigiata*
	Dasycladales	Dasycladaceae	*Bornetella sphaerica Neomeris annulata*
	Ulvales	Ulvaceae	*Ulva prolifera Ulva armoricana Ulva fenestrate Ulva lactuca Ulva rigida Enteromorpha prolifera*
	Ulotrichales	Monostromataceae	*Monostroma grevillei*
		Ulotrichaceae	*Spongomorpha aeruginosa*
	Prasiolales	Prasiolaceae	*Prasiola calophylla*

Table 1.
List of important seaweed species.

found in shallow waters and withstand deep water and low light conditions. It has been reported that 59 species of seaweeds found in Indonesian coastal region, 15 of the species are capable of stimulating germination, development, growth, and production of rice and legume crops [10]. *Halimeda opuntia*, *Caulerpa racemosa*, *Gracilaria edulis*, and *Chaetomorpha crassa* grow widely and abundantly in stagnant water in the mangrove hinterlands of Indian and Pacific Ocean. The seaweeds easily grow, survive, and monopolize in stagnant water where the salinity and temperature greatly fluctuate. *Hypnea musciformis* is found in Atlantic, Pacific, and Indian Ocean and has the widest geographical distribution. *Mastophora rosea, Mastophora pacifica, and Mastophora multistrata* were found in bluish purple color and distributed in Pacific Ocean and Indian Ocean. *Sarcodiotheca chordalis* was reported in the Mediterranean Sea and Atlantic Ocean from Morocco to Southwest England. *Cheilosporum cultratum* and *Caulerpa lentillifera* have been newly found in near the islands of Indian Ocean. Totally 59 species of seaweeds were recorded from the Indian Ocean which included six species of Scytasiphonaceae and Gracilariaceae, five species of Ulvaceae, four species of Cladophoraceae and Halimedaceae, three species of Dictyotaceae, two species of Caulerpaceae, Rhodomelaceae, Corallinaceae, Florideophyceae, Sargassaceae, Cystocloniaceae Gelidiellaceae, and Oscillatoriaceae, and one species of Bangiophyceae, Lithophyllaceae, Codiaceae, Cystocloniaceae, Characea, Valoniaceae, Boodleaceae, Siphonocladaceae, Solieriaceae, Galaxauraceae, Liagoraceae, Sphacelariaceae, Monostromataceae, Ulvophyceae, and Lessoniaceae [11]. *Kappaphycus alvarezii* is a tropical seaweed being cultivated in southeast coastline of India since more than a decade for the extraction of thickening agent called kappa-carrageenan using traditional farming systems.

3. Significance of seaweed in agriculture

Liquid products based on marine algae were introduced in 1950 and now are successfully used worldwide. The seaweed concentrates are admistered to crops as root dips, soil drenches, and foliar sprays. Seaweed-derived biostimulant extracts used as foliar sprays have gained importance for multiple crops including various grasses, cereals, pulses, and vegetables [12]. Seaweed extract application as foliar spray is a common practice to boost production in many commercial crops. Seaweed extract enhances tolerance against environmental stresses and challenges and enhances nutrient uptake from soil by plants [13]. Foliar application of seaweed biostimulants offers a quicker method of supplying nutrients to higher plants alternate to soil application method [14]. This may as a result of active nutrient uptake through stomatal pores instead of cuticular uptake [15]. The mechanism by which seaweed biostimulants affect cellular metabolism is based on the physiological action of macro- and microelements, amino acids, vitamins, and substances. The seaweed biostimulant enhances meristematic growth, translocation of photosynthates, enzyme activation, cell elongation, and cell stability [16]. Seaweed extract enhances chlorophyll content by increasing the biogenesis of chloroplasts and reducing chlorophyll degradation, which was due to the upregulated genes associated with photosynthesis, cell metabolism, stress response, and S and N metabolism in *Brassica napus* L. [17]. The seaweed when applied to crops as foliar spray can accelerate the rate of cell division and elongation. Seaweed extract of the brown algae *Sargassum heterophyllum* contain cytokinin, while *Ascophyllum nodosum* contain the growth hormone Indole-3-acetic acid. Seaweed supplies macro- and micronutrients, has liming properties, and increases phosphorus availability, the phycocolloids improve soil structure, increases the water retention and cation-exchange capacities of soil, binds metals, boosts biological activity, and improves plant resistance to aggressive biotic and abiotic agents [18].

4. Seaweed extract on biochemical characters in crops

The bioactive compounds (not yet fully elucidated) present in *Ascophyllum nodosum* extracts when applied to stressed plants have reduced the deleterious effects of drought stress by regulating a series of sequential molecular, cellular, and physiological responses including the modulation of several genes, resulting in an

Parameters	*Ascophyllum nodosum*	*Laminaria digitata*
Type	Brown	Brown
Water (%)	70–80	73–92
Ash (%)	15–26	21–37
Alginic acid (%)	15–35	20–50
Laminaren (%)	0–15	0–20
Mannitol (%)	5–10	4–20
Fucoidan (%)	4–15	2–6
Carbohydrate (%)	10	1–3
Protein (%)	5–10	8–20
Fat (%)	2–8	1–3
Tannins (%)	2–12	1–2
Potassium (%)	2–3	1.3–3.8
Sodium (%)	3–4	0.9–2.2
Magnesium (%)	0.5–0.9	0.5–0.8
Iodine (%)	0.01–0.1	0.3–1.3

Table 2.
Chemical composition (in percentage) of the seaweeds Ascophylum nodosum and Laminaria digitata.

accumulation of various osmolytes, an improved antioxidant system, and enhanced gaseous exchange through stomatal regulation (**Table 2**). Priming wheat seeds with the extracts of *U. linza* or *C. officinalis* promoted almost physiological parameters like chlorophyll, carotenoids, sugars, proteins, and lipids. Seaweed extracts with high magnesium and mineral content tended to increase leaf total chlorophyll and carotenoids concentration [19]. The increased protein and sugar contents due to seaweed priming could be attributed to absorption of majority of the major elements in these extracts, particularly magnesium, which could have activated chlorophyll synthesis and, as a result, improved photosynthetic rates [20]. Foliar spray of seaweeds extract was effective in improving wheat performance by enhancing compatible osmolytes and antioxidant compounds and enhancing variation among non-coding chloroplast DNA (cpDNA) regions trnL intron and psbA-tnH as a response to water deficit.

5. Physiological effect of seaweeds extracts application in mitigating abiotic stress

Application of seaweed extracts as natural regulators has had promising benefits, such as enhanced crop productivity and plant vigor to withstand adverse environmental influences. Plants, being sessile, are relentlessly challenged by a variety of environmental stresses that limit their growth and productivity. Due to the complex metabolic pathways involved in stress tolerance, limited success has been achieved in generating stress-tolerant crops through genetic engineering. Extract of *Ascophyllum Nodosum* application has shown to assist soybean plants endure severe drought conditions by regulating leaf temperature, turgor, and several stress-responsive genes [21]. Mild salinity stress causes physiological drought in plants, impairing cell–water relations, inhibiting cell expansion, and, consequently, reducing growth rate. Soil salinity is a global problem, affecting over 800 million hectares of land, resulting in massive impacts on agricultural productivity [22]. Long-term exposure

Figure 1.
Common seaweeds species in Ocean (A), Rhodophyceae (Red algae), (B) Phaeophyceae (Brown algae), and (C) Ulvophyceae (Green algae).

to high salinity causes ionic stress by disturbing the homeostasis of intracellular ions, which results in membrane dysfunction and attenuation of metabolic activity and secondary effects, inhibiting growth and inducing cell death [23]. Mild salinity stress induces physiological drought in plants, impairing cell–water relations, inhibiting cell expansion, and consequently, reducing growth rate. Salinity induces both ionic and osmotic stresses, thus reducing plant growth and productivity [24]. *Ulva prolifera* extract reduced the oxidative damage caused by drought not only by activating the antioxidative system but also by providing essential hormones and minerals for wheat growth [25]. Drought stress is one of the most severe abiotic stresses which have negative impact on plant growth, crop production, and numerous metabolic processes. Application of seaweed extract in mitigating water stress and adverse effects is highly important for plant production (**Figure 1**).

6. Effect of seaweed extract on crop development and yield

Seaweed extract (15 percent) was sprayed as foliar application to rice crop during vegetative and generative stages. The results shown that extracts of *Sargassum calophyllum*, *Sargassum polyceratum*, *Sargassum vulgare*, *Sargassu*m *aquifoliu*m, *Sargassum polycystum, Hydroclathrus clathratus, Hydroclathrus minutes, Ulva fasciata, Ulva ferticulata, Padina pavonica, Gracilaria edulis, Chaetomorpha crassa, Turbinaria ornata*, and *Turbinaria murayana* were able to promote growth of rice plants. *Hydroclathrus clathratus* extract enhanced crop growth and production of rice crop. This phenomenon may be due to the presence of active compounds and micro- and macronutrients in the extract of seaweeds which can stimulate plant growth (**Figure 2**). Various species of marine algae found in nature contain organic compounds which activities resemble the activity of cytokinins, auxins, and gibberellins. The organic compounds were able to stimulate growth as a result of enhancement of protein synthesis, cell division, and mobilization of nutrients needed to boost growth of rice crop [9].

Seaweed extract supplement of *Sargassum horneri* was applied at 5 percent to fertilizer (50 percent) as basal and top dressing increased the rice yield slightly and, especially, the amount of fertilizer applied was saved by 50% in rice crop [26]. Three foliar sprays of *Kappaphycus alvarizii* (K sap) and *Gracilaria edulis* (G sap) when applied at the doses of 2.5, 5.0, 7.5, and 15.0 percent (v/v) coupled with water spray as a control at different phases of the maize crop significantly enhanced grain yield significantly by 18.5 percent and 26.0 percent. The enhancement in yield was correlated to increase in the number of rows in cob, cob length, and 100 grain weight and improved in nutrient uptake [27]. The substantial increase in plant height, root nodules, root volume, number of branches at harvest, dry matter production, chlorophyll content of the leaves, grain yield, grain nitrogen percentage, and grain protein content was significantly recorded by soaking seed in 0.1% seaweed extract solution for 30 minutes + foliar application of seaweed extract (0.25%) twice on 25 DAS and 35 DAS in green gram (single spray). The existence of micro- and macronutrients, trace elements, humic acid, amino acids, plant growth hormones, vitamins, antibiotics, carbohydrates, metabolite boosters, and other organic compounds in seaweed extract significantly enhanced the growth, yield, and quality traits of green gram [28].

Spraying of *Kappaphycus alvarezii* sap (5.0%) in tomato crop significantly increased the root length, shoot length, and yield over (control) plants sprayed with water. The result was due to macro- and microelements as well as growth-promoting components like cytokinin present in seaweed. There was significant increase and

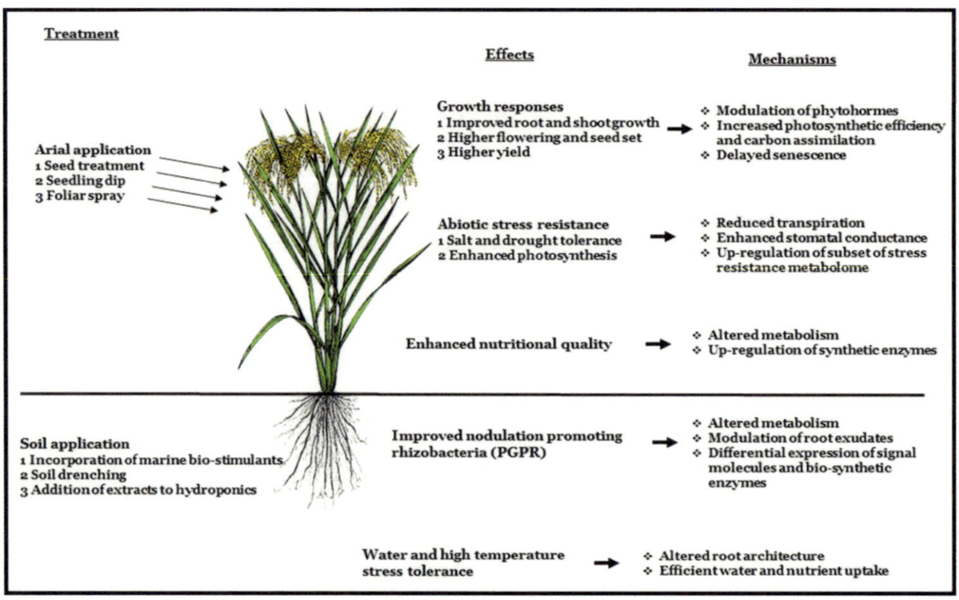

Figure 2.
Illustration of physiological effects of elicited seaweed biostimulants and possible mechanisms of bioactivity.

enhancement in chlorophyll and b content in leaves over control where *Kappaphycus alvarezii* was applied (5.0% sap). Enhanced leaf chlorophyll concentration of treated plants with seaweed biostimulant was due to the presence of betaines [29]. Spraying rice crop with seaweed extracts at rate of 2000 ppm led to a significant enhancement in the weight of 100 grains and produced the highest significant values of grains yield per plant under salt stress conditions. The superiority was due to the benefits of seaweed components providing a remarkable source of bioactive substances such as macro- and micronutrients, essential amino acids, fatty acids, vitamins, cytokinins, and auxins like growth-promoting substances influencing plant growth and production by altering cellular metabolism in treated plants. In rice, higher plant height, dry matter production, yield attributes such as number of grains per panicle, panicle length, thousand grain weight, number of productive tillers per m^2, grain yield, and straw yield were recorded when seaweed extract was applied to the soil at a rate of 12.5 kg/ha along with a foliar spray of seaweed liquid [30]. The application of liquid seaweed extracts improved plant growth and yield characteristics and enhanced soil quality in foxtail millet [31].

Seaweed liquid biostimulant of *Kappaphycus alvarezii* sap when applied as foliar application in rice crop boosted the crop growth characteristics such as plant height, number of leaves plant^{-1}, number of tillers plant^{-1}, and grain yield over recommended dose of fertilizers [32]. Increase in yield is due to higher number of panicles through improved canopy establishment, photosynthesis, translocation of assimilates, better interception of light, and markedly decrease in inter plant competition for soil nutrients and solar energy. Seaweed extracts applied on onion grown under water stress significantly enhanced N, P, and K uptake by 116, 113, and 93% compared to the unsprayed plants [33]. Agronomic characters and yield components of rice in algalized plots increased compared to the treatment with recommended dose of urea [34].

In vitro mass propagation using hypocotyls and leaf disk explants of brinjal (*Solanum melongena* L.) cultivar Pusa Purple Long, the influence of seaweed liquid extracts (SLEs) prepared from *Gracilaria salicornia*, *Padina gymnospora*, *Padina boergesenii*, and *Gelidiella acerosa* on plant growth was studied. The rate of shoot and root induction and the percentage of seed germination were both markedly improved by the use of seaweed liquid extract in MS medium. Plant growth can be promoted by seaweed extracts contain stimulating compounds. Seaweed extracts contain high endogenous levels of micro- and macronutrients, vitamins, cytokinins, auxins, amino acids, gibberellins, and organic acids to enhance the growth of crop plants [35].

Seaweed foliar spray sap of *Kappaphycus alvarezii* and *Gracilara edulis* at five percent and above concentration boosted grain yield attributes and yield of rice 5.4 to 19.4 percent higher as compared to recommended dose of fertilizers. Soaking of rice seeds in lower concentration (2.5 and 5%) of saps extracted from seaweed improved the germination and seedling vigor of rice [36]. *Padina gymnospora*, *Gracilaria edulis*, and *Ulva fasciata* aqueous extracts applied as a seaweed biofertilizer to promote seed germination of *Capsicum annuum* illustrated higher root and shoot length, mean germination time, germination index, germination percent, seedling vigor index, germination energy, total protein content, total phenol content, and antioxidant inhibition percentage [37]. Seaweed extract of *Sargassum denticulatum* (2 percent) foliar application enhanced all growth and yield parameters and more accumulation of the organic solutes in leaves of water-stressed plants. Application of seaweed biostimulant/extract enhanced seedling growth of wheat plants during growth period [38]. The influence of three red marine algae species (*Laurencia obtusa*, *Corallina elongata*, and *Jania rubens*) was assessed as biostimulant to enhance the growth of maize (*Zea mays* L.) plants. Application of mixture of *Laurencia obtusa* and *Corallina elongata* enhanced plant fresh and dry weight in maize crop [39]. Seaweed sap foliar application (*Kappaphycus alvarezii* and *Sargassum denticulatum*) at 10 percent at 20–40 DAS and 40–60 DAS revealed maximum accumulate growth rate (AGR), crop growth rate (CGR), and relative growth rate (RGR) over 5 percent and 7.5 percent along with foliar application water spray in maize. Seaweed extract mainly contains amino acids like betaines and sterols which enhance the photosynthetic activity, N metabolism, and protein synthesis which boost corn production and also the availability of growth regulators to extract especially auxins and cytokinins which are responsible for cell enlargement and inter-nodal elongation and thereby increase the vegetative growth [40]. Application of seaweed extract as foliar at 2 percent concentration enhanced all growth and yield parameters and more accumulation of the organic solutes in leaves of water-stressed crop plants. Low concentrations application of *Ascophyllum nodosum* extracts on the ground or on the foliage of rice caused an increase in chlorophyll content in the leaves. The decrease in chlorophyll degradation, which might be partially attributed to betaine from seaweed extract, resulted in an increase in chlorophyll content and photosynthetic rate [41].

Application of seaweed extract at the rates of 15% K sap or 15% G sap with 100% RDF at 35, 45, and 60 days after transplanting (three distinct intervals) over the required dose of fertilizer enhanced grain yield by 11.80 percent and 9.52 percent, respectively. The growth, yield attributes, grain yield, quality, and chlorophyll content of rice were significantly influenced with foliar spraying seaweed extract on the foliage. In accordance with the results, spraying seaweed extract at 15 percent K sap with 100 percent recommended dose of fertilizer resulted in substantially higher growth, yield attributes, and chlorophyll content over spraying seaweed extract at 15 percent G sap with 100 percent recommended dose of fertilizer. Yield enhancements in seaweed-treated plants are responsible to be associated with the hormonal components present in the extracts, especially cytokinins [42]. The most significant values of grain yield per plant were achieved after spraying wheat plants with seaweed extract at a concentration of 2000 ppm under salt stress conditions. The superiority was induced by the advantages of seaweed components, which function as a predominant source of bioactive compounds like macro- and micronutrients, essential fatty acids, amino acids, vitamins, cytokinins, and auxins, substances that affect cellular metabolism in treated plants and hence promote growth and productivity [43]. Application of seaweed biostimulant at a dose of 2 ml L^{-1} demonstrated positive impact on the growth, development, yield, and essential oil contents of coriander plant [44].

7. Effect of application of seaweed biostimulant on nutrient uptake

Rhizosphere microorganisms are conducive to soil nutrient cycling for plant growth. Rhizosphere microorganisms contribute to the release of organic acids, amino acids, carbohydrates, and secondary metabolism products in rice fields. Seaweed extracts have been reported to enhance tolerance to environment stress, boost the growth and yield of plants, and promote nutrients availability and nutrients uptake from the soil. Since seaweed extract is the source of organic matter and nutrients, so the utilization of seaweed extract as biofertilizer could balance the N, P, and K deficiencies in paddy soils. Root of rice is a hidden organ in soil that mediates critical functions, involving the uptake and storage of water and nutrients [45]. Rhizosphere microbes are critical for plant growth and biogeochemical cycles and are closely related to cultivated crop, fertilizer applications, and plant residue input [46]. *Sargassum horneri* extract, alginate, can chelate with major cations of Na^+ , $Ca_2{}^+$, $Mg_2{}^+$, and K^+ to form aggregate with richer nutrients, improve the crumb structure and capillary activity of soil pores, and, finally, boost soil microbial activity [47]. Therefore, seaweed extract supplement can potentially be a strategy of enhancing the amount of nitrogen, phosphorous, and potassium contents in soil. Application of seaweed extract on plant is capable of enhancing nutrient concentrations in the leaves, through implication of growth hormone in the process of nutrients absorption and movements in a plant, thus enhancing the weight of the rice plant. The utilization of seaweed-based extracts has reduced the levels of nitrogen, phosphorus, and potassium fertilizers and also induced the seed germination and growth parameters strongly than chemical fertilizers in rice crop [10]. *Ecklonia maxima* seaweed extract is most effective in neutral pH soil, and it can be used to promote plant growth under low pH and water stress conditions [47]. Application of algae mixture of *Corallina elongate, Laurencia obtusa,* and *Jania rubens* caused increase in phosphorus content and nitrogen content in maize plant [39]. Decomposed seaweed, as an organic matter, generally improved soil physico-chemical properties, water holding capacity, and microbial activity and also protected plant against unfavorable environmental conditions such as extreme temperatures and water stress [48].

Application of seaweed concentrate led to significantly increase of K, Mg, and Ca concentrations in the leaves of lettuce plants which received adequate supply of nutrients but had meager effect on nutrients-stressed plants. Moreover, the components, nitrogen, phosphorus, and magnesium, were significantly increased when seaweed extract applied through foliar application. Seaweed application in meager quantities has the effect on several metabolic processes and improves plant growth and development through enhancing of photosynthesis, endogenous hormones, nutrients uptake, and protein synthesis as well as with relatively improved ability for increasing available micronutrients in the soil [49]. Foliar applications of *Ascophyllum nodosum* seaweed extract reinvigorated the plants to utilize with soil mineral N and other available nutrients more efficiently which enhanced increased grain potassium uptake and increased in wheat plant yields by 25 percent [50]. Due to increased membrane permeability of roots, leaves, and stoma cells, as well as hormone-like activities of seaweed extract through their involvement in cell respiration, photosynthesis, and enzymatic processes, the application of seaweed extract boosted the uptake of copper. Application of 5 percent of *Kappaphycus alvarezii* sap enhanced the uptake of N, P K, S, Ca, K, and Mg in leaves of rice crop. Seaweed extract supplement could impact the bacterial community in tillering and heading stages in rice crop, and hence α-diversity of rhizosphere bacteria in the heading stage improved substantially [24]. In the rice environment, enhanced soil available nutrients of N (260 kg/ha), P (42 kg/ha), K (180 kg/ha), Ca (27.7 meq/100g), Mg (5.5 meq/100g), S (18.2 mg/kg), Zn (1.17 ppm), Fe (33.82 ppm), Cu(1.61 ppm), and Mn (18.97 ppm) were observed after applying seaweed extract to the soil at a rate of 25 kg [28]. *Ulva linza* and *Calendula officinalis* contain macronutrients such as N, P, K, Ca, Mg, and some trace elements, as well as growth regulators, amino acids, and antioxidants which are excellent natural fertilizers [51].

Elevation in uptake of K, Ca, Mg, and Cu with application of seaweed extract of *Kappaphycus alvarizii* was due to enhanced membrane permeability of roots, leaves and stoma cells, and hormone-like activities of seaweed extract through their implication in cell respiration, photosynthesis, and enzymatic reactions [27]. Application of seaweed at the rate of 15 percent *Kappaphycus alvarezii* in black gram in sandy loam soil of the red and lateritic belt of West Bengal resulted in higher availability and absorption of inorganic elements like Ca, Na, K, Mg, N, Zn, and Cu [52].

8. Conclusion

Abiotic stress such as drought, high soil salinity, and temperature affect and limit crop productivity worldwide. Biostimulant seaweed application reduces the need for fertilizers and enhances nutritive efficiency, abiotic stress tolerance, grain yield, and plant quality traits. Application of minimal concentration of seaweed extract as a priming agent enhances plant growth, physiological attributes, and molecular traits. Application of seaweed extracts of *Kappaphycus alvarezii*, *Ascophyllum nodosum*, *Sargassum calophyllum*, *Sargassum aquifolium*, *Sargassum polycistum*, *Hydroclathrus clathratus*, *Turbinaria ornata*, and *Turbinaria murayana* are able to stimulate vegetative growth and increase yield of rice plants. The use of seaweed biostimulants in low quantities has effect on several metabolic processes and improves plant growth and development through the increase of photosynthesis, endogenous hormones, nutrients uptake, and protein synthesis as well as with relatively higher capacity for increasing available micronutrients in the soil. In order to boost crop growth parameters like plant height, leaf area index, dry matter production, and SPAD reading (chlorophyll content), as well as yield parameters like the number of grains per panicle, panicle length, and number of products, seaweed extract *Kappaphycus alvarezii* was applied to rice crop (Oryza sativa L.) at a rate of 12.5 kg/ha in the soil and 0.5% twice at the tillering and panicle initiation stages. The rice yield was also increased by 18–20% over the recommended fertilizer dosage. However, the application of seaweed extract in soil at the rate of 25 kg/ha improved the available nutrients like N, P, K, Ca, Mg, and S and micronutrients. Hence, seaweed has a substantially greater potential to boost bioavailable macro- and micronutrients to the rice crop.

Conflict of interest

The authors declare no conflict of interest.

Author details

Melekote Nagabhushan Arun[1*], Rapolu Mahender Kumar[1], Sailaja Nori[2], Banugu Sreedevi[1], Guntupalli Padmavathi[3], Pallakonda Revathi[4], Neha Pathak[5], Dayyala Srinivas[1] and Raman Meenakshi Sundaram[6]

1 Department of Agronomy, Indian Council of Agricultural Research - Indian Institute of Rice Research, Hyderabad, Telangana, India

2 Sea6 Energy Private Limited, Bangalore, Karnataka, India

3 Department of Plant Breeding, Indian Council of Agricultural Research - Indian Institute of Rice Research, Hyderabad, Telangana, India

4 Department of Agronomy, PJTSU, Hyderabad, Telangana, India

5 Department of Botany, Aligarh Muslim University, Aligarh, Utter Pradesh, India

6 Department of Biotechnology, Indian Council of Agricultural Research - Indian Institute of Rice Research, Hyderabad, Telangana, India

*Address all correspondence to: arun_tulasi2011@yahoo.in

References

[1] Naomi F, Lewis Z. Rice: Importance for global nutrition. Journal of Nutritional Science and Vitaminology. 2019;**65**:S2-S3

[2] Food and Agriculture Organization. FAO STAT; 2020. Available from: http://www.fao.org/faostat/en/#data/QC/visualize

[3] Yakhin OI, Lubyanov AA, Yakhin IA. Physiological activity and efficiency of application of biostimulants. Agrokhimya. 2016;**6**:72-94

[4] Barbier M, Charrier B, Araujo R, Holdt SL, Jacquemin B, Rebours C. In: Barbier M, Charrier B, editors. Pegasus-Phycomorph European Guidelines for a Sustainable Aquaculture of Seaweeds. Roscoff, France: Station Biologique de Roscoff; 2019

[5] Nasmia RE, Masyahoro A, Putera FHA, Natsir S. The utilization of seaweed-based liquid organic fertilizer to stimulate *Gracilaria verrucosa* growth and quality. International Journal of Environmental Science and Technology. 2021;**18**:1637-1644

[6] Karthikeyan K, Shanmugam M. The effect of potassium-rich biostimulant from seaweed kappaphycus alvarezii on yield and quality of cane and cane juice of sugarcane var. Co 86032 under plantation and ratoon crops. Journal of Applied Phycology. 2017;**29**:3245

[7] El Arroussi H, Benhima R, Elbaouchi A, Sijilmassi B, El Mernissi N, Aafsar A. Dunaliella salina exopolysaccharides: A promising biostimulant for salt stress tolerance in tomato (*Solanum lycopersicum*). Journal of Applied Phycology. 2018;**30**:2929-2941

[8] Ali O, Ramsubhag A, Jayaraman J. Biostimulant properties of seaweed extracts in plants: Implications towards sustainable crop production. Plants. 2021;**10**(3):531

[9] Bhushan S, Mehra R, Rani P, Bast F. DbIndAlgae: An on-line resource of marine algae from India identified on the basis of molecular and morphological features. Current Science. 2016;**111**:723-726

[10] Sunarpi AJ, Kurnianingsih R, Julisaniah NI, Nikmatullah A. Effect of seaweed extracts on growth and yield of rice plants. Bioscience. 2010;**2**(2):73-77

[11] Devipriya S, Jeyapraba L. Checklist and distribution of seaweeds from Manapadu region- Southeast coast of India. The International journal of analytical and experimental modal analysis. 2020;**12**(8):2456-2459

[12] Karthikeyan K, Munisamy S. Bio-stimulant of seaweed source as an organic alimentative to Bellary onion: Bulb yield and Pyruvic acid levels. American Journal of Agricultural Research. 2016;**1**:1-9

[13] Yadav S, Arvind V, Nepalia V. Effect of phosphorus, sulphur and seaweed sap on growth, yield and nutrient uptake of chickpea (*Cicer arietinum* L.). Research on Crops. 2016;**17**:496

[14] Nedumaran T, Arulchandran D. Seaweeds: A promising source for sustainable development. In: Thangavel P, Sridevi G, editors. Environmental Sustainability. India: Springer; 2015. pp. 65-88

[15] Fernandez V, Brown PH. From plant surface to plant metabolism: The

uncertain fate of foliar-applied nutrients. Frontiers in Plant Science. 2013;**4**:289

[16] Pramanick B, Brahmachari K, Ghosh A. Efficacy of *Kappaphycus* and *Gracilaria* sap on growth and yield improvement of sesame in new alluvial soil. Journal of Crop and Weed. 2014;**10**(1):77-81

[17] Serenella N, Pizzeghello D, Schiavon M, Ertani A. Plant biostimulants: Physiological responses induced by protein hydrolyzed-based products and humic substances in plant metabolism. Scientia Agricola. 2016;**73**(1):18-23

[18] Begum M, Bordoloi BC, Ojha DDSNJ. Role of seaweed extract on growth, yield and quality of some agricultural crops: A review. Agricultural Reviews. 2018;**39**(4):321-326

[19] Mohy El-Din SM. Utilization of seaweed extracts as bio-fertilizers to stimulate the growth of wheat seedlings. Egypt Journal of Expermintal Biology. 2015;**11**:31-39

[20] Castellanos-Barriga LG, Santacruz-Ruvalcaba F, Hernández-Carmona G, Ramírez-Briones E, Hernández-Herrera RM. Effect of seaweed liquid extracts from *Ulva lactuca* on seedling growth of mung bean (*Vigna radiata*). Journal of Applied Phycology. 2017;**29**:2479-2488

[21] Shukla PS, Mantin EG, Adil M, Bajpai S, Critchley AT, Prithiviraj B. *Ascophyllum nodosum* – Based biostimulants: Sustainablea applications in agriculture for the stimulation of plant growth, stress tolerance and disease management. Frontiers in Plant Science. 2019;**10**:655

[22] Ferchichi S, Hessini K, Dell'Aversana E, D'Amelia L, Woodrow P, Ciarmiello LF. *Hordeum vulgare* and *Hordeum maritimum* respond to extended salinity stress displaying different temporal accumulation pattern of metabolites. Functional Plant Biology. 2018;**45**:1096

[23] Shukla PS, Gupta K, Agarwal P, Jha B, Agarwal PK. Overexpression of a novel SbMYB15 from *Salicornia brachiata* confers salinity and dehydration tolerance by reduced oxidative damage and improved photosynthesis in transgenic tobacco. Planta. 2015;**242**:1291-1308

[24] Agarwal PK, Shukla PS, Gupta K, Jha B. Bioengineering for salinity tolerance in plants: State of the art. Molecular Biotechnology. 2013;**54**:102-123

[25] Kasim WA, Saad-Allah K, Hamouda M. Seed priming with extracts of two seaweeds alleviates the physiological and molecular impacts of salinity stress on radish (*Raphanus sativus*). International Journal of Agriculture and Biology. 2016;**18**:653-660

[26] Chen C-L, Song W-L, Lin S, Qin S, Ren C-G, Yang J-C, et al. Effect of seaweed extract supplement on rice rhizosphere bacterial community in tillering and heading stages. Agronomy. 2022;**12**:1-15

[27] Basavaraja PK, Yogendra ND, Zodape ST, Prakash R, Ghosh A. Effect of seaweed sap as foliar spray on growth and yield of hybrid maize. Journal of Plant Nutrition. 2018;**41**(14):1-14

[28] Ishwarya R, Vaseeharan B, Kalyani S, Banumathi B, Govindarajan M, Alharbi NS, et al. Facile green synthesis of zinc oxide nanoparticles using *Ulva lactuca* seaweed extract and evaluation of their photocatalytic, antibiofilm and insecticidal activity. Journal of

Photochemistry and Photobiology. 2018;**178**:249-258

[29] Zodape ST, Gupta A, Bhandari SC, Rawat US, Chaudhary DR, Eswaran K, et al. Foliar application of seaweed sap as biostimulant for enhancement of yieldand quality of tomato (*Lycopersicon esculentum* Mill.). Journal of Scientific & Industrial Research. 2011;**70**:215-219

[30] Deepana P, Sathiya Bama K, Santhy P, Sivasankari DT. Effect of seaweed extract on rice (Oryza sativa var. ADT53) productivity and soil fertility in Cauvery delta zone of Tamil Nadu India. Journal of Applied and Natural Science. 2021;**13**(3):1111-1120

[31] Rathinapriya P, Lakkakula S, Subramani P, Rameshkumar R, Manokaran B, Kasinathan R, et al. Effects of liquid seaweed extracts in improving the agronomic performance of foxtail millet. Journal of Plant Nutrition. 2020;**43**(19):2857-2875

[32] Arun MN, Mahender Kumar R, Nori S, Singh A, Tuti MD, Srinivas D, et al. Effect of seaweed extract as biostimulant on crop growth and yield in rice (*Oryza sativa* L.) under transplanted condition. Journal of Rice Research. 2019;**12**(2):45-51

[33] Almaroai Yaser A, Eissa MA. Role of marine algae extracts in water stress resistance of onion under semiarid conditions. Journal of Soil Science and Plant Nutrition. 2022;**20**:1092-1101

[34] Thamida Begum ZN, Mandal R, Saiful I. Effect of cyanobacterial biofertilizer on the growth and yield components of two HYV of rice. Journal of Algal Biomass Utilization. 2011;**2**(1):1-9

[35] Lakkakula S, Periyasamy R, Arokia SR, Antony CS, Subramani P, Ramakrishnan R, et al. Somatic embryogenesis and regeneration using *Gracilaria edulis* and *Padina boergesenii* seaweed liquid extracts and genetic fidelity in finger millet (*Eleusine coracana*). Journal of Applied Phycoogy. 2015;**27**:993-1002

[36] Layek J, Das A, Idapuganti RG, Sarkar D, Ghosh A, Zodape ST, et al. Seaweed extract as organic bio-stimulant improves productivity and quality of rice in eastern Himalayas. Journal of Applied Phycology. 2018;**30**:547-558

[37] Rajan MSA, Thriunavukkarasu R, Joseph J, Aruni W. Effect of seaweed on seed germination and biochemical constituents of *Capscicum annum*. Biocatalysis and Agricultural Biotechnology. 2020;**29**:101761

[38] Ali AH, Said EM, Abdelgawad ZA. The role of seaweed extract on improvement drought tolerance of wheat revealed by osmoprotectants and DNA (cpDNA) markers. Brazilian Journal of Botany. 2022;**1**:1-11

[39] Safinaz AF, Ragaa AH. Effect of some red marine algae as biofertilizers on growth of maize (Zea mayz L.) plants. International Food Research Journal. 2013;**20**(4):1629-1632

[40] Meshram MR, Dawson J, Singh S, Rina L. Effect of seaweed sap foliar application on AGR, CGR and RGR of maize (*Zea mays* L.) cultivars. Biological Forum – An International Journal. 2022;**14**(2):111-116

[41] Shukla PS, Mantin EG, Adil M, Bajpai S, Critchley AT, Prithiviraj B. *Ascophyllumnodosum*-basedbiostimulants: Sustainable applications in agriculture for the stimulation of plant growth, stress tolerance, and disease. Management. Frontiers in Plant Science. 2019;**10**:655

[42] Leindah Devi N, Mani S. Effect of Seaweed Saps *Kappaphycus alvarezii* and *Gracilaria articulata* on growth, yield and quality of rice. Indian Journal of Science and Technology. 2015;**8**(19):1-6

[43] Salim BBM, Abdel-Rassoul M. Effect of foliar applications of seaweed extract, potassium nitrate and potassium silicate on growth yield and some biochemical constitutuents of wheat plants under salt stress. Journal of Biological Chemistry and Environmental Science. 2016;**11**:371-391

[44] Tursun AO. Effect of foliar application of seaweed (organic fertilizer) on yield, essential oil and chemical composition of coriander. PLoS One. 2022;**17**(6):e0269067

[45] Pascale A, Proietti S, Pantelides IS, Stringlis IA. Modulation of the root microbiome by plant molecules: The basis for targeted disease suppression and plant growth promotion. Frontiers in Plant Science. 2019;**10**:174

[46] Li H, Su JQ, Yang XR, Zhu YG. Distinct rhizosphere effect on active and total bacterial communities in paddy soils. Science of the Total Environment. 2019;**649**:422-430

[47] Yang SH, Seo J, Koo Y. Alginate and fucoidan changes the bacterial community in different directions and the alginate or fucoidan degrading bacteria isolated from paddy soil promotes the plant growth. Achieves of Microbiology. 2021;**203**:5183-5192

[48] Di Stasio E, Rouphael Y, Colla G, Raimondi G, Giordano M, Pannico A, et al. The influence of *Ecklonia maxima* seaweed extract on growth, photosynthetic activity and mineral composition of Brassica rapa L. subsp sylvestris under nutrient stress conditions. European Journal of Horticultural Science. 2017;**82**:286-293

[49] Anderson G. Seaweed extract shows improved fruit quality at McLaren Vale vineyard trial. Australian and New Zealand Grapegrower and Winemaker. 2009;**548**:17-22

[50] Stamatiadis S, Evangelou L, Yvin J-C, Tsadilas C, Mina JMG, Cruz F. Responses of winter wheat to *Ascophyllum nodosum* (L.) Le Jol. extract application under the effect of N fertilization and water supply. Journal of Applied Phycology. 2021;**33**:1873-1882

[51] Hamouda MM, Saad-Allah KM, Gad D. Potential of seaweed extract on growth, physiological, cytological and biochemical parameters of wheat (*Triticum aestivum* L.) seedlings. Journal of Soil Science and Plant Nutrition. 2022;**22**:1818-1831

[52] Ghosh A, Tanmoy S, Malik GC, Banerjee M, Ghosh A. Effect of seaweed extracts on the growth, yield and nutrient uptake of black gram (*Vigna mungo* L.) in the red and lateritic belt of West Bengal. International Journal of Chemical Studies. 2020;**8**(3):799-802

Chapter 3

Seaweed *Kappaphycus alvarezii* Cultivation for Seagrass Ecosystem Conservation

Rajuddin Syamsuddin

Abstract

The physical characteristics of the seagrass ecosystem indicate that the shallow sea waters are ideal for seaweed cultivation. The rapid development of *Kappaphycus alvarezii* seaweed cultivation in coastal areas of Indonesia should not cause damage to the seagrass ecosystem. The study on the cultivation method of *K. alvarezii* seaweed in three seagrass ecosystems in Indonesia showed high growth rates, biomass production and carrageenan content due to high nutrient concentrations and high water clarity as well as the current optimal conditions in the cultivation environment. *K.alvarezii* cultivated in seagrass ecosystems prevents the damaging effects of UV-B radiation on those ecosystems. The right cultivation method applied is the off-bottom method.

Keywords: biomass, carrageenan, growth rate, light intensity, *Kappaphyus alvarezii*, nutrients, seagrass, seaweed sustainability, UV-B radiattion

1. Introduction

As many as 25,742 hectares of seagrass spread out in the coastal areas of Indonesia. The seagrass ecosystem is one of the most productive (high organic productivity) marine ecosystems and with high biodiversity. It is supporting fishery production, namely as feeding ground, spawning ground, nursery ground, as well as shelter from various predators and from the hot sun for various fish species.

Seagrass ecosystem are flowering plant (Angiospermae) vegetation that are formed by one or more species with high or rare density that are able to fully adapt in high salinity waters, which cover a coastal shallow marine zone. These physical characteristics of the seagrass ecosystem seems to be the ideal location for seaweed cultivation.

Seaweed cultivation and business in Indonesia are experiencing rapid development and plays an important role in improving people's welfare. However, seaweed cultivation in coastal areas should be carried out in an environmentally friendly manner, without damaging the seagrass ecosystem. *K. alvarezii* is a species of red seaweed (Rhodophyceae) which changed its name from *Eucheuma cottonii* to *Kappaphycus alvarezii* on the basis of the kappa-carrageenan it contains [1], is one of economically important red (macro) algae species that has a high demand on the world market.

Based on the above considerations, a series of studies were conducted in the period 2011–2020 to find the right *K. alvarezii* cultivation method in the seagrass ecosystem that shows high growth rates, biomass production, and carrageenan content, and evaluates its impact on the condition of the seagrass ecoosystem. The research was conducted in three seagrass ecosystems in coastal areas of Indonesia, namely in the coastal waters of Tanakeke, Mappakasunggu District, and the coastal waters of Laikang Bay, Mangarabombang District, both are administrative areas of Takalar Regency, South Sulawesi Province. The water depths of both waters of the research location were 500 cm and 250 m, respectively, during high tide. Both have the same current pattern with current velocities ranging from 4.0–8.5 m/second outside the seagrass ecosystem. Another location is the grass ecosystem in Karampuang Island, Mamuju Regency, West Sulawesi Province with a depth of 400 cm during high tide. These three seagrass ecosystems are in one waters, namely the Makassar Strait.

The cultivation method tested in the seagrass area in Takalar Regency was a longline system with three methods, namely the *floating surface longline method*, the *off bottom method*, and the *bottom method*. *The floating method* was placed 20 cm below the sea level, *the off-bottom method* was placed 50 cm above the bottom sediment of the seagrass ecosystem, and *the bottom method* was placed right on the surface of the bottom sediments. While the method used in Mamuju Regency was only *the floating surface longline* method. Seaweed seeds of *K. alvarezii* as much as 100 g/bundle were tied to nylon ropes with a distance of 50 cm between ties and between ropes. Cultivation time was 42–45 days.

Seaweed growth was expressed as daily growth rate and calculated according to the formula suggested by Fortes [2]; Dawes *et al.* [3]; Mtolera *et al.* [4]; Hurtado *et al.* [5]; Munoz *et al.* [6]; Hayashi *et al.* [7]. Some water quality parameters were measured at the research site, including: light intensity falling on the sea surface using a Lux-meter, salinity using a handrefractometer, temperature using a thermometer, pH using a pH meter, current velocity with a current meter, and free CO2 with titration method [8]. Other chemical parameters were analyzed at the Water Quality Laboratory, Department of Fisheries, Faculty of Marine and Fishery Sciences, Hasanuddin University, Makassar, Indonesia, consisting of nitrate using the sulfuric acid method [9], ammonium and orthophosphate measured using a spectrophotometer [8]. Carrageenan levels were analyzed by extraction of dried seaweed with 0.5% KOH solution at a temperature of 90–95°C.

2. Seaweed growth

Seagrasses absorb nutrients directly from the water column through their leaves and in sediments by their roots [10], their sediments store a lot of accumulated organic matter, 60–80% of which is found in the roots (rhizomes) of seagrass plants, the rest is in the form of stem and leaf fragments of the plants, tissues fragment of aquatic animals, and others. Therefore, the sediment of the seagrass ecosystem is a good habitat for decomposing bacteria. Through the bacteria and various benthic organisms activities, the organic matter are decomposed and releases carbon, nitrogen, and phosphorus and other mineral nutrients. The carbon production of seagrass ecosystems is quite high, ranging from 900 to 4650 $gC/m^2/year$ [11] which is a source of high concentration of dissolved free CO2.

As a submerged soil, ammonium ($NH4^+$)is the main product of mineralization (decomposition) of nitrogen-containing organic compounds in sedimentary soils of seagrass ecosystems [12]. Ammonium is the dominant form of nitrogen in seagrass ecosystems [13, 14]. Some of the $NH4^+$ is oxidized by *Nitrosomonas* bacteria to nitrite ($NO2^-$) in the initial phase of the Nitrification process, and the next phase is the oxidation of $NO2^-$ by *Nitrobacter* to $NO3^-$. In addition, under aerobic conditions at the bottom of the seagrass ecosystem that occur periodically at low tide, some ammonium is oxidized to nitrate, so that nitrate is also a form of nitrogen that is also quite abundant in seagrass ecosystems, and is also absorbed by seaweed. Therefore, as a source of nutrients, the sediment continuously supplies carbon, ammonium and nitrate to the above water layer through the processes of turbulence and diffusion. Aerobic bacteria that works around the roots of seagrass plants convert phosphorus that is in a complex state (bound with Fe ions so that it cannot be absorbed by seaweed) into dissolved phosphorus which can be absorbed by seaweed [15].

In the seagrass ecosystem in Takalar Regency, the concentrations of free carbon dioxide, orthophosphate, nitrate, and ammonium ranged from 6.40–67.92 ppm, 0.064–0.599 ppm, 0.015–0.65 ppm and 0.047–0.704 ppm, respectively down water column (at all cultivation methods). This value is suitable for seaweed cultivation. The N and P content in seagrass sediments with fine soil particles is higher compared to environments with sandy sediments [16] outside the seagrass ecosystem [15]. These nutrients are stored in high concentrations in sedimentary clay particles of seagrass ecosystems [17], then released into the water column for further absorption by *K. alvarezii* [18].

Nitrogen is an important nutrient in the process of cell division, for seaweed growth, but is most often reported limiting the growth of seaweed [19, 20]. Ammonium (NH4), which is a form of nitrogen ion (N) with a higher concentration than the concentration of nitrate (NO3), allows high growth rates of seaweed that was cultivated in seagrass ecosystems. The absorption rate of ammonium by seaweed is higher than that of nitrate [21], because ammonium is a form of nitrogen that enters the metabolic process. Nitrate is reduced to ammonium before combining into organic compounds [22]. Several previous studies have shown that ammonium is a form of N that is directly utilized by plants in protein biosynthesis. In plant cells, nitrate is reduced by the nitrate reductase enzyme to ammonium, the form of N which then combines with organic compounds in the stimulates vegetative growth (thallus) [23].

As a constituent of protoplasm, phosphorus plays a role in reducing plant abortion (stopping organ growth), plays a role in the formation of meristem tissue (tissue consisting of actively dividing cells), stimulates cell division and repairs damaged tissue [24]. This chemical is found in high concentrations in the water column of seagrass ecosystems.

The growth rate of seaweed in the floating method was 1.13–1.53%/day, lower than the off-bottom method (1.34–1.72%/day) on seagrass in the coastal waters of Laikang Bay. The low growth of seaweed using the floating method is probably due to the high exposure to UV-B radiation in the surface water layer. In the surface water layer, photosynthetic pigments (chlorophyll a and carotenoids) are damaged by excessive light intensity and by the damaging effect of ultraviolet UV-B radiation, known as photoinhibition, photodamage (damage by light) and photooxidation (oxidation by light) [25]. Light intensity measured at the water surface in the two seagrass ecosystems ranged from 3700 to 4100 Lux, exceeding the light intensity of 600 Lux for the maximum growth rate of seaweed. The optimal light intensity for seaweed growth ranged from 333 to 1000 Lux [26]. This low level is also caused by the *Warburg Effect* phenomenon due to aerobic (oxic) conditions, which is a very high concentration of dissolved oxygen in water beyond the optimal concentration due to the diffusion of oxygen from the atmosphere to the water column in the surface layer combined with dissolved oxygen produced during the photosynthesis of the seaweed itself that was cultivated with floating method. The Warburg effect is the phenomenon of competition between carbon dioxide and oxygen on the reactive site of RUBP carboxylase (an enzyme that binds carbon dioxide gas in the dark reaction of photosynthesis), causing only a small amount of free carbon dioxide to be reduced to carbohydrates by the enzyme in the pentose phosphate pathway of the dark (reaction) phase of photosynthesis [24, 27].

With the same off-bottom method, *K. alvarezii* cultivated in the seagrass ecosystem of Laikang Bay showed a lower growth rate (1.34–1.72%/day) compared to 2.20–2.54%/day those cultivated in the Tanakeke seagrass ecosystem. With the same cultivation method, the quantity of sunlight (the same range of 3700–4100 Lux the light intensity falling on the water surface of both waters) which is dispersed, attenuated, and reflected as light travels through the water column at low depths (200 cm in Laikang Bay Seagrass) just a little so that the intensity of sunlight that penetrates the column and exposes the tallus of seaweed is still quite high. Seaweed growth at lower water depths of the seagrass ecosystem, Laikang Bay may be caused by light intensity that exceeds 1000 Lux [28]. While the light hitting the thallus which is positioned at a greater depth (in Tanakeke seagrass ecosystem) the intensity has been greatly reduced (become lower and at least approximately in optimal range) by dispersion, attenuation, and reflection.

The difference in current velocity in these two locations also causes differences in the growth rate of seaweed in these two locations. The current velocity at the bottom of the seagrass ecosystem in Tanakeke is relatively faster (54–56, cm/second) which has an impact on better nutrient absorption by seaweed compared to nutrient absorption of seaweed in the Laikang seagrass ecosystem with a slower current velocity (20–40 cm/sec). The relatively faster current speed causes nutrient uptake relatively more efficient due to the thinner *boundary layer* (the space or layer between the water carrying nutrients with the seaweed thallus). However, current velocities of 20–40 cm/sec and 54–56, cm/second are considered optimal for nutrient absorption. With this method, photosynthesis and absorption of nutrients (CO2, NH4, NO3 and PO4), both via passive ion absorption through osmosis and diffusion processes and active absorption take place effectively. In addition, seaweed is also physiologically protected from the Warburg Effect at 450 cm water depth.

The low growth rate (1.01–1.27%/day) of seaweed using the bottom method in Takalar Regency can be caused by bacterial activity that breaks down the seaweed thallus which is in direct contact with the bottom of the waters.

High growth *of K.alvarezii* (2.26–2.42%) was only obtained in Mamuju District at the age of 30 days after growth. Beyond the 30th to 45th day of cultivation, there was a decrease in growth which was only 0.28–0.56%/day on the 45th day due to frequent rains at the cultivation location which caused the condition (range) of water quality parameters no longer optimal. The decrease in transparency (high water turbidity) and salinity, as well as the presence of pollutants in the form of waste from residential areas carried by surface run-off by rainfall events are the main causes of the low growth of seaweed. In addition, the influence of river flow that empties into the coastal waters of the cultivation site is the main factor causing the low growth rate of *K. alvarezii*. These factors were also thought to cause some seaweed thallus to be susceptible to a seaweed disease known as ice-ice which causes seaweed death.

The main cause of ice-ice disease is the condition of extreme abiotic factors that exceed the tolerance limit of seaweed [29] such as exposure to very high light intensity, low water salinity, and nutritional deficiencies that cause seaweed to become susceptible to bacterial infections including *Pseudomonas*, viruses and fungi. The initial symptom of this pathogen is in the form of white spots on the thallus, then it changes color to pale, the texture is easily crushed, rots and finally the thallus falls out of the clump.

3. Biomass production

In the seagrass ecosystem in Takalar Regency, the highest seaweed biomass production (18.2–30.83 g/clump) was obtained with the off-bottom method, in line with the growth rate which was also the highest at that particular method, and low in the bottom method (12.4–18.52 g/clump). Nitrogen available in high concentrations in seagrass ecosystems causes high biomass production of *K. alvarezii* cultivated in seagrass ecosystems. This nutrient determines the productivity of algae [14], is very important in the synthesis of chlorophyll a [30] which plays an important role in the photosynthesis process that produces biomass.

4. Carrageenan content

In the process of decomposing organic matter at the bottom of the water, bacteria utilize O2 and release CO2 into the water column which is then absorbed by seaweed in the process of photosynthesis which produces carbohydrates, followed by a secondary product in the form of carrageenan. Carrageenan (phycocolloid) is a natural additive that is widely used in various industries, especially the food, pharmaceutical and cosmetic industries. *K. alvarezii* seaweed cultivated in seagrass ecosystems in Takalar Regency produced a fairly high content of carrageenan (39.9–44.8%) in the seagrass ecosystem of Laikang waters, and higher (40.73–50.16%) those grown in the Tanakeke seagrass ecosystem. The carrageenan content is in line with the growth rate because the factors that affect growth are also factors that affect the carrageenan content. Several previous studies [31, 32], also showed that the carrageenan content of K. *alvarezii* was positively correlated with its growth rate. Those that grown in Mamuju Regency also have high levels of carrageenan ranging from 44.95–49.15%, although with a lower growth rate when compared to those that grow in Takalar Regency. These levels of carrageenan exceeds the quality standard set by FAO for industrial raw materials, which is 40%. The carrageenan content of *K. alvarezii* was relatively higher compared to the carrageenan content of 40.7% recorded by Munoz *et al.*, [6], 31.2–38.1% by Hurtado *et al.*, [33], and 30.57–36.93% by Syahrul *et al.*, [34] all of which were cultivated outside the seagrass ecosystem.

Variations in carrageenan content are influenced by several factors, including cultivation location and climate [35]. The high content of carrageenan *K. alvarezii* cultivated in seagrass ecosystems is caused by the concentration of dissolved ammonium in seagrass ecosystem waters [13, 14]. Luning [36] suggested that nutrients (including nitrogen) in sufficient quantities can increase the synthesis of polysaccharides (carrageenan).

Since ammonium instead of nitrate is a form of nitrogen that is directly utilized in protein (amino acid) biosynthesis in plant cells, the abundant availability of ammonium in the seagrass ecosystem limits the energy (NADPH) that will be used in reducing nitrate to ammonium in protein biosynthesis in algal cells and more energy is used to reduce carbon dioxide in the synthesis of organic compounds (including carrageenan) in the dark reactions of photosynthesis [37] *K. alvarezii*. According to Sahoo and Ohno [19], the concentration of ammonium absorbed by seaweed causes high levels of carrageenan in seaweed.

5. Water temperature, pH and salinity

In addition to nutrients in the form of nitrate, ammonium and orthophosphate, seaweed growth is also influenced by environmental conditions such as salinity, temperature, and sunlight [38]. Water temperature, pH and salinity are environmental factors that affect the metabolism, growth and production of seaweed. Temperature affects the metabolic rate (photosynthesis and respiration) of algae [36, 39]. The temperature of the seagrass ecosystem in Takalar Regency in all vertical water layers is around 29–30°C. Meanwhile in Mamuju Regency the temperature ranged from 29.7–32°C respectively. In general, the temperature required for seaweed growth ranges from 20 to 30°C [36]. *K. alvarezii* can grow at a temperature of 25–29°C [40]. In the tropics the growth rate and biomass production of *K. alvarezii* is high in the temperature range of 25–30°C [41].

pH is a chemical factor that determines the availability of nutrients to be absorbed by plants so that it affects the growth of the seaweed. The range of water pH measured in seagrass ecosystems in Laikang, Tanake, and Mamuju Bays was 7.5–7.8, 7.2–8.0, and 7.5–8.4. The pH range indicates that the three seagrass ecosystems are classified as waters with high productivity. This pH range is optimal for the growth and development of *K. alvarezii* based on Trono's [42] statement that seaweed can live in a pH range of 7.5–8.4. In this pH range, the decomposition of organic matter that accumulates at the bottom of the seagrass ecosystem takes place more quickly, and immediately releases various nutrients needed by seaweed. $HPO4^{3-}$ is the predominant P ion that is absorbed at pH over 8.

The salinity of the seagrass ecosystem in Takalar Regency in all vertical water layers is 30–31 ppt. Meanwhile in Mamuju Regency the salinity ranged from 29.7–32 ppt. This salinity range is suitable for the growth of *K. alvarezii* seaweed in the tropics. The salinity range of 18–35 ppt is suitable for seaweed growth [6], and very good in the range of 22–30 ppt [43]. Good salinity range for the growth of *Eucheuma* sp. is 30–35 ppt [44]. Doty [45] stated that the desired salinity of *K. alvarezii* ranged from 29 to 34 ppt. Sulu *et al.* [40] stated that 30 ppt salinity was best for the growth of *K. alvarezii*.

6. *Kappaphcus alvarezii* cultivation and sustainability of seaweed ecosystems

Based on the above facts (data from studies conducted in seagrass meadows in Takalar and Mamuju districts as well as supporting academic references), *K. alvarezii* seaweed can grow well side by side with seagrass vegetation, even with high carrageenan content, without having to cut down the seagrass vegetation. There is no competition in meeting nutrient needs between the two because seagrass plants absorb more nutrients from sediment, while seaweed only takes nutrients that are soluble in water. Even, seagrass vegetation supports the growth of seaweed through nutrient cycles from the sediment to the water layer where the seaweed grows [15]. These nutrients are released into the water column after the decomposition of the leaves and other parts of the seagrass.

With stem and leaf morphology resembling the stems and leaves of land grass and reed plants, as well as their root system that propagates in and on the surface of the bottom sediment, seagrass acts as a sediment trap and stabilizes the bottom sediment [46], so it is not easy to stirred by the movement of water (current and waves), dampening the waves and currents that cause the water mass to become calm and clear (high transparency) in this ecosystem. This ecosystem has the characteristics of low current velocity, as a storehouse of mineral nutrients which are recycled from the sediment to the water column above it. This condition is really needed by seaweed for its growth and development. Therefore, high daily growth rate, biomass production and carrageenan content of *K. alvarezii* were obtained using off-bottom cultivation method in seagrass ecosystems. With the availability of natural food in the form of leaves of seagrass plants, epiphytic algae and animals attached to the sediment substrate, leaves, stems and roots of seagrass, the attack of herbivorous fish on seaweed can be reduced.

Until now, there is no indication of any negative impact of seaweed cultivation on seagrass growth, especially for *Enhalus acoroides* seagrass. Most seagrass species can survive in low light intensity by means of physiological as well as morphological, or the and anatomical structure adaptations of the leaves [47]. Growth is mainly influenced by water quality parameters. During the cultivation of seaweed in these three seagrass ecosystems, the seaweed farmers activities did not cause physical damage to the seagrass plants. Seagrass degradation is most likely to occur if the plant is trampled by the cultivator. However, when doing seaweed cultivation activities in the seagrass ecosystem, this does not happen because the series of activities ranging from attaching ropes to harvesting are all carried out by boat without stepping on seagrass plants at the bottom of the waters.

The seagrass meadows has high biodiversity and is one of the marine ecosystems with high organic productivity, which supports food chains and food webs in the sea, both based on herbivorous and detrivorous chains. Through the decomposition process of detritus and other organic matter, these waters are a natural food habitat for benthic animals and filter feeder invertebrates that consume detritus particles. With soft bottom sediments, seagrass ecosystems become niches for various benthic animals (zoobenthos), including polychaeta, crustaceans (shrimp, prawns, crabs), sea urchins, mollusks, and sea cucumbers. Important economic fish species that are often found foraging in this ecosystem include rabbit fish (*Siganus canaliculatus, S. fuscescens, S. guttatus), Mugil* spp., grouper, *Epinehelus* spp., *Neonniphon sammara, Scarrus* sp., *Lethrinus harak, L. .orantus, Lutjanus kasmira, Pomacantus semicirculatus, Calatus spinidens*, and *Parupenous berberinus*. Some of these fish are either as a seasonal resident, visitors, or occasional residents in this ecosystem. Other marine organisms are sea horses (*Hyppocampus* spp), turtles, and dugongs.

As a shallow marine ecosystem, seagrass ccosystems are very sensitive to the adverse effects of ultraviolet radiation, especially UV-B (wavelength 280–320 nm) which exposes and penetrates down the water column of seagrass ecosystem. With thin leaf anatomy, seagrass plants are sensitive to UV radiation, so that in the future, this radiation will have a negative impact on the sustainability of this ecosystem [48].

K. alvarezii, a species of red algae (Rhodophyta) known as a group of algae with high red, orange and yellow pigments in its tissues. These pigments are carotenoids and their derivatives, such as lutein, loroxanthin and siphonaxanthin content. Carotenoids are generally found in cell membrane systems where one of the main functions of these compounds is related to photosynthesis. The carotenoid content of *K. alvarezii* that we recorded in our study on the cultivation of this species in two districts of Takalar Regency was 0.02–0.03% in the dry season when light intensity was high enough, and 0.06–0.17% in the rainy season where relatively lower light intensity. This explains that the carotenoid synthesis is high under conditions of low light intensity, and low under high light intensity. Carotenoids are accessory pigments that function to capture sunlight energy at certain wavelengths beyond the wavelengths that can be captured by chlorophyll and other photosynthetic pigments [49]. The formed carotenoids then act as protecting compounds against the destruction effect of ultraviolet UV-B radiation on the seagrass plants, known as photoinhibition, photooxydation and photodamage [50–52]. Through cultivation in an environment with high nutrient concentrations, the carotenoid content of *K. alvarezii* cultivated in seagrass ecosystems is also high.

As a macroalgae, *K. alvarezii* contains mycosporine-like amino acid compounds (MAA), the compound that gives seaweed species the ability to tolerate the effects of UV-B radiation. Under certain conditions, through the pigmentation mechanism in which the synthesis of carotenoid pigments increases, and with a thick thallus, with a multicellular layer, as well as morphological adaptations where the *K. alvarezii* clumps become denser when there is an increase in light intensity, cultivating this algae species in the seagrass ecosystem provides shade for seagrass from the UV-B radiation, and can physiologically reduce or inhibit the negative effects of the radiation on seagrass plant tissues. Thus, through the off-bottom method of seaweed in this seagrass ecosystem, it is possible to minimize the destructive impact of UV-B radiation before exposing seagrass plants, and recover the damage from the effects of UV-B radiation. Thus, this cultivation can maintain the sustainability of the seagrass ecosystem as a provider of ecological and socioeconomic services.

7. Conclusions

Based on the discussion, it can be concluded that *K. alvarezii* seaweed can grow well with high carrageenan content in the seagrass ecosystem, without having to clear seagrass plants. By cultivating seaweed in the seagrass ecosystem, the seagrass vegetation is protected by the seaweed clumps from the negative effects of ultraviolet radiation which can damage the seagrass vegetation in the long term. With the off-bottom cultivation method and 100 grams the initial weight of the seedling with a tie spacing of 50 cm, the cultivation of *K. alvarezii* in the seagrass ecosystem did not inhibit the growth and development of seagrass vegetation.

DOI: http://dx.doi.org/10.5772/intechopen.106762

Author details

Rajuddin Syamsuddin
Faculty of Marine Science and Fisheries, Fisheries Department, Hasanuddin University, Makassar, Indonesia

*Address all correspondence to: rajuddin_syamsuddin@yahoo.com

References

[1] Doty MS. *Kappaphycus alvarezii* (Gigartinales, Rhodophyceae) from Malaysia. In: Abbot JA, Norris La Jolla J, editors. Taxonomy of Economic Seaweed with Reference to some Pacific and Carribean Species. California: California Sea Grant College Program; 1985

[2] Fortes ETG. Introduction to the Seaweed. Their Characteristics and Economic Importance. Report in Training Course of *Gracillaria* Algae. Manila Philippines: Up-South China Sea Project; 1981

[3] Dawes CJ, Lluisma AO, Trono GC. Conal propagation of *Eucheuma denticulatum* and *Kappaphycus alvarezii* in the Philippines Journal of Applied Physiology, 1993;**6**:21-24

[4] Mtolera MSP, Collen J, Pedersen M, Semesi AK. Destructive hydrogen peroxide production in *Eucheuma denticulatum* (Rhodophyta) stress during the caused by elevated pH, high light intensities and competition with other species. European Journal of Phycology. 1995;**30**:289-297

[5] Hurtado AQ, Agbayan RF, Sanares R, M.R.de Castro-Mallare. The seasonality and economic feasibility of cultivating *Kappaphycusnalvarezii* in Panagatan, cays, Caluya, antique, Philippines. Aquaculture. 2001;**199**:295-310

[6] Munoz J, Freile-Pelegrin Y, Robledo D. Marculture of *Kappaphycus alvarezii* (Rhodophyta, Solieriaceae) color strains in tropical waters of Yucatan, Mexico. Aquaculture. 2004;**239**:161-177

[7] Hayashi L, Paula EJD, Chow F. Growth rate and carrageenan analysis in four strains of *Kappaphycus alvarezii* (Rhodophyta, Gigartinales) farmed in the subtropical water of Sao Paulo State, Brasil. Journal of Applied Phycology. 2007;**19**:393-399

[8] Stricland JDH, Parson CJ. A Practiced Handbook of Seawater Analysis. Bd of Canada Ottawa, Canada: Fish. Res; 1970. p. 310

[9] APHA. Standard Methods for the Examination of Water and Wastewater. 20th ed. American Public Health Association, American Water Works Association and Water Environmental Federation, Washington DC; 1998

[10] Tomasick T, Mah AJ, Nontji A, Moosa MK. The ecology of Indonesian sea. Part II Periplus sedition, (HK) ltd. Singapore. 1997

[11] Bengen DG. Ekosistem andsSumberdaya Alam Pesisir Laut. Pusat Kajian Kajian Sumberdaya Pesisir and Lautan. Bogor: Institut Pertanian Bogor; 2001

[12] Ponnamperuma FN. The Physico-Chemical Properties of Submerged Soils in Relation to Fertility. IRPS ed. No.5. IRRI, Los Banos, Philippines; 1977

[13] McCarthy JJ, Taylor WR, Taft JL. Nitrogenous Nutrition of the Plankton In the Chasepeake Bay. 1. Nutrient Availability and Phytoplankton Preferences. Limnol and Oceanogr (USA);1997;**22**(6):996-1011

[14] Prescott GW. The Algae; A Review. Boston: Houghton; Mifflin Co; 1968. p. 430

[15] Moriarty DJW, Boom PI. Interactions of seagrasses with sediment and water. In: Larkum AWD, McComb AJ, Shepherd SA, editors.

Biology of Seagrasses. Amsterdam-New York: Elsevier; 1989. pp. 501-535

[16] Brady CN. The Nature and Propertis of Soils. New York: MacMilan Publishing. Co; 1974. berpasir

[17] Golterman HC. Chemistry. In: Whitton BA, editor. River Ecology. Studies in Ecology. Oxford, London, Edinburgh, Melbourne: Blackwell Scientific Publications. Vol. 2. 1975. pp. 81-105

[18] Gruending GK. Ecology of epipelic communities in Modia Lake. British Columbia Journal of Phycology. 1975;7(3):239-249

[19] Sahoo D, Ohno M. Culture of *Kappaphycus alvarezii* in deep seawater and nitrogen enriched medium. Bulletin of Marine Science Fish Kochi University. 2003;**22**:89-96

[20] Hanisak. Cited by (in) Sahoo and Ohno, 2003. Culture of Kappaphycus alvarezii in deep seawater and Nitrogen enriched medium. Bulletin of Marine Science Fish Kochi University. 1983;**22**:89-96

[21] Wheeler WN, Srivastata LM. Seasonal nitrate physiology of *Macrocystis integrifolia* Bory. Journal of Experimental Marine Biology and Ecology. 1984;**76**:35-50

[22] Syret PJ. Nitrogen assmimilation. In: Lewin RA, editor. 1962, Physiology and Biochemistry of Algae. New York: Academic Press; 1962. pp. 171-188

[23] Peni DKS, Anggarwulan E. Pertumbuhan, Kadar Klorofil, Karotenoid, Saponin, Aktivitas Nitrat Reduktase Anting-Anting (*Acalyph Indica*) Pada Konsentrasi Asam Giberelat (GA3) Yang Berbeda. Sebelas Maret. Surakarta: Jurusan Biologi FMIPA, Univ; 2003

[24] Salisbury FK, Rose C. Plant Physiology. Belman, California: Wadsworth Publ. Co. Inc.; 1969. p. 764

[25] Gleen EP, Doty MS. Photosynthesis and respiration of the tropical red seaweed. *Eucheuma striatum* (Tambalang and Eikhorn varieties) and *E. denticulatum*. Aquatic Botany. 1981;**10**:353-364

[26] Wood WF. Photoadaptive responses of the tropical red algae *Euchema striatum* Schmitz (Gigartinales) to ultraviolet radiation. Aquatic Botany. 1989;**33**:44-51

[27] Noggle GR, Fritz GJ. Intyroductory Plant Physiology. 2nd.ed. ed. Englewood Cliffs, New Jersey: Prentice-hall, Inc.; 1983. p. 627

[28] Mairh OP, Soe-Htun U, Ohno M. Culture of *Eucheuma striatum* (Rodophyta, Solieraceae) in sub tropical of Shikoku, Japan. Botanica Marina. 1986;**29**:185-191

[29] Trono GC. Eucheuma farming in the Philippines. Quezon City, Philippines: U. P. natural science research Centre; 1974. pp. 10-15. Uan, J. 1990, Kiribati

[30] Lawlor DW. Photosynthesis. 2nd ed. Longman Group UK Limited; London. 1993. pp. 9-23

[31] Hurtado AQ, D.A., Yanque, K. Tibulos, and A. T. Critchley. Use of Acadian marine plant extra powder of Ascophyllum nodusum in tissue culture of *Kappaphycus alvarezii*. Journal of Applied Phycology. 2009;**21**:633-639

[32] Naguit MRA, Tisera WL, Lanioso A. Growth Performance and Carrageenan Yield of *Kappaphycus Alvarezii* (Doty) and *Eucheuma Denticulatum* (Burman) Collins et Harvey, Farmed In Bais Bay, Negros Oriental and Olingan, Dipolog City. Journal the Threshold. 2009;**IV**:38-51

[33] Hurtado AQ, Crithchley AT, Trespoey A, Bleicher-Lhonneur G. Growth and carrageenan quality of *Kappaphycus striatum* var. *Sacol* grown at different stocking densities, duration of culture and depth. Journal of Applied Phycology. 2008;**20**:551-555

[34] Syahrul MYK, Thana D. Pengaruh Berbagai Metode Penanaman Terhadap Pertumbuhan And produksi Rumput Laut *Eucheuma Spinosum*. Laporan Penelitian Strategis nasional Batch IV. Makassar: Universitas Hasanuddin; 2009

[35] Nurjannah. Prospek pemanfaatan rumput laut. Seminar Diversifikasi Rumput Laut. Makalah pada Seminar Rumput Laut tanggal 3 Mei 2003. Fakultas Perikanan and Ilmu Kelautan. Bogor: Institut Pertanian Bogor; 2003

[36] Luning K. Seaweeds; their Environment, Biogeogrpaphy and Ecophysology. New York: John Wiley; 1990. p. 527

[37] Syamsuddin R. Growth of phytoplankton *Isochrysis galbana* under various nitrogen sources and concentrations. Torani, Bull. IK UNHAS. 1997;**7**(1):23-28

[38] Thirumaran G, Anantharaman P. Daily growth rate of field farming seaweed *Kappaphycus alvarezii* (Doty) Doty ex P. silva in Vellar estuary. World Journal of Fish and Marine Sciences. 2009;**1**(3):144-153

[39] Gessner F. Temperature in plants. In: Kinne O, editor. Marine Ecology. Vol. 1. 1970. pp. 363-406

[40] Sulu R, Kumar L, Hay C, Pickering T. Kappaphycus seaweed in the Pacific; review of introduction and field testing proposed quarant Aquaculture Technical Papers/Pickering)ISSN 1683-7568 ine protocols. In: The Institutes of Marine Resources. The University of the South Pacific; 2003

[41] Trono GC, Ohno M d. Seasonality in the biomass production of the *Eucheuma* strains in northern Bohol, Philippines. In: Umezaki I, editor. Scientific Survey of Marine Algae and their Resources in the Philippine Islands. Technical Report, Monbusho International Scientific Research Program, Japan. 1989. pp. 71-80

[42] Trono GC. Present Status of Culture of Tropical Agarophytes. Nivercity of the Philipphines. Metro. Manila: Quenzon City; 1989

[43] Raikar SV, Lima dan M, Fujita Y. Effect of temperature, salinity and light intensity on the growth of *Gracilaria* spp. (Gracilariales; Rhodophyta), from Japan, Malaysia and India. Indian Journal of Marine Science. 2001;**30**:98-104

[44] Dawes CJ. Marine Botany. New York: John Wiley And Sons University of South Florida; 1981. p. 268

[45] Doty MS. The production and use of Eucheuma in case studies of seven commercial seaweed research. In: Doty MS, Caddy JF, Santillices B, editors. FAO Technical Paper No. 281. Rome; 1987

[46] Den Hartog C. Seagrasses and seagrass ecosystem; an appraisal of the research approach. In: Aquatic Botany. Amsterdam: Elsevier Sci. Publ.; 1979. pp. 105-117

[47] Collier CJ, M, Waycott and AG.. Ospina. Responses of four indo-West Pacific seagrass species to shading. Marine Pollution Bulletin. 2012;**65**:342-354

[48] Dawson SP, Dennison WC. Effects of ultraviolet and photosynthetically active radiation on five seagrass species. Marine Biology. 1996;**125**:629-638

[49] Cunningham FX, Jr., Gantt E. One ring or two? Determination of ring number in carotenoids by lycopene ε-cyclases. Proceedings of the National Academy of Sciences of the United States of America. 2001;**98**:2905-2910

[50] Mendoza WG, Ganzon-Fortes ET, Villanueva RD, Romero JB, Montano MNE. Tissue age as factor affecting carrageenan quantity in farmed Kappaphycus striatum. Bot.Mar. 2006;**49**:57-64

[51] Mtolera MSP, Buriyo AS. Studies on Tanzanian Hypneaceae: Seasonal variation in content and quality of kappa-carrageenan from Hypnea musciformis (Gigartinales : Rhodophyta). Western Indian Ocean Journal of Marine Science. 2004;**3**(1):43-49

[52] Glenn EP, MS Doty. Photosnthesis and respiration of the tropical red seaweed Eucheuma striatu (Tambalang and elkhorn varieties) and E. denticculatum. Aquatic Botany, 1981;**10**:353-364

Chapter 4

Marine Bivalves' Ecological Roles and Humans-Environmental Interactions to Achieve Sustainable Aquatic Ecosystems

Andreia Filipa Mesquita, Fernando José Mendes Gonçalves and Ana Marta Mendes Gonçalves

Abstract

Bivalve species have a key role at the ecosystem level and a very interesting economic value. Globally, bivalve production is higher than 15 million tons. Thus, this work intends to highlight the economic value of these organisms, but mostly highlights the potential of this resource for water management and water quality improvement, and thus to the sustainability of aquatic systems, which gives them a particular interest. These organisms are under anthropogenic pressures becoming crucial to preserve aquatic systems and their communities, namely bivalve communities, and water quality by reducing pollution. UN Sustainable Development Goals (SDGs) highlight the main actions to reduce humans' footprint and to create globally a model to guarantee human security, to protect the environment and water quality and to combat climate changes. To achieve the UN SDGs, bivalves may have a high importance for sustainability and preservation of freshwater and marine systems (SDG 14), and for water management (SDG 6), due to their ability to improve the water quality by reduction of pollution. This work aims to highlight the main ecological roles of marine bivalves and the human actions that will contribute to achieve sustainable aquatic systems, and so the SDG 6 and SDG 14 by 2030.

Keywords: bivalves, food resources, water quality and management, sustainability, ecological roles

1. Introduction

Water is a valuable resource to human beings; however nowadays, 2.2 billion people continue without access to safe drinking water [1].

Freshwater ecosystems provide us several services, such as transport, natural purification, protection against floods, irrigation, and they are characterized by a high biodiversity of habitats. Unfortunately, the discrepancy between developing and developed countries is considerable. In developed countries, 3.5% of the land is covered by freshwater, compared to just 1.4% in developing countries [1]. Despite the importance of freshwater, this only includes less than 3% of all water in the world. By other side, oceans comprise about 97% of all water in the planet [2] and have an important ecological role, as habitat to many animal and plant species, of all sizes, being essential efforts to the conservation and preservation of these systems. In addition to the ecological importance, these ecosystems have a great value to the human development, since they provide many resources, including economic activities as fishing, communications, tourism, and recreation activities. Marine vegetation is responsible for the production of about 50% of atmospheric oxygen; mangroves, saltmarshes, and seagrasses have a key role as natural sinks of carbon; coastal habitats present high importance to the protection of habitats, communities and businesses against storm, and wave damage; and humans ascribe great esthetic, cultural and spiritual values from ocean ecosystems [3]. So, due to the great potential of these ecosystems, human populations have developed their activities and inhabited near the sea, with 67% of human population living less than 400 km from the sea [4].

Notwithstanding the great value of the aquatic systems to many species and to human well-being, these environments have been neglected and subjected to many pressures, such as resource overexploitation (e.g., overfishing) and input of pollutants from human activities (e.g., industrial and agricultural practices, shipping, beach activities). Therefore, these pressures result in dangerous effects to the entire ecosystems, including consequences to the biotic (e.g., changes in the structure and function of the communities) and to the abiotic (e.g., effects on water quality and sediment composition) elements. Considering the importance of each species and the ecosystem as a whole, it is essential to work for the preservation and conservation of these ecosystems, reducing the impacts of anthropogenic activities and improving water quality. So this work attempts to highlight the human role in the sustainable management of freshwater and marine aquatic systems, focusing on bivalve communities, due to their ecological and economic importance.

2. SDG 6 and SDG 14

Sustainable Development Goal 6 intends to achieve an equitable access to drinking water and sanitation for all people, through several targets. It includes the elimination or reduction of the pollutant inputs to aquatic systems, encouragement of the cooperation between countries for the implementation of integrated water resource management, and assurance of the protection and recovery of water related ecosystems (e.g., wetlands, lakes, and rivers). SDG 14 is focused on marine ecosystems' preservation and conservation. This is achieved through measures to prevent and reduce marine pollution, including land activities, to minimize ocean acidification and its consequences (e.g., decreased carbonate ions' interference with shell formation, affecting the survival of bivalves, plankton, and coral reefs) and by promoting the sustainable use of marine resources through economic benefit assessment.

Considering these two SDGs and their targets, it is crucial human beings and different sectors take actions to achieve the 2030 Agenda main goals. Several efforts are essential, such as effective policies to control the impact on aquatic systems, both from contaminant discharges and from resource overexploitation. Moreover, spreading the knowledge of environmental issues and involving people in environmental activities may also be fundamental to the awareness of the non-scientific communities, making easier the implementation of policies and the achievement of the goals. Ocean Literacy and Environmental Education programs for school students and the public in general may contribute to promote knowledge, awareness, and governmental action.

The critical position of bivalve species in the ecosystems, acting as a link between primary producers and secondary consumers, and their ability in filtering and cleaning the water and in accumulating great amounts of nutrients and contaminants is well known. Their role in carbon and nutrient cycles is fundamental to achieve clean water and ecosystem health. Thus, a compilation of bivalves' importance and the role of anthropogenic impacts on aquatic systems, for achieving SDGs 6 and 14 are provided in the section below.

3. Ecological and economical role of bivalves species

Bivalves (Bivalvia) are a large class, with almost 15,000 species (MolluscaBase 2021). These organisms are found worldwide in marine and freshwater ecosystems and they are reported as playing a key role at the ecological level, due to their large filtration ability. These organisms are filter feeders and filter plankton and other suspension particulates, contributing to the water cleaning and light penetration. They also have an essential part in the ecosystem hydrodynamics as well as in nutrient and carbon cycles [5]. Their high filtration ability and potential for use in bioremediation processes [6] make them potential key elements for the improvement and restoration of water quality. Moreover, they are a food source to many species, from invertebrates to fishes, birds, and mammals [7], acting as a link between primary producers and secondary consumers. Briefly, they can establish and change habitats and affect the trophic webs, directly and indirectly [8]. By this, bivalve species preservation is an important contribution to the ecosystem's equilibrium and resource sustainability, being a key element to achieve SDGs 14 and 6.

Despite the great ecological importance of these organisms, bivalve species also have a great economic value to humans, since they are used not only as ornamental objects but also as a food and protein source. For instance, in 2017, 40,000 tons of global oyster were imported, and about 170,000 tons of clams were exported and 180,000 tons imported, with China as the main exporter whereas Japan and Republic of Korea as the main importers [9]. Scallop import prices in the French market were US$15/kg for frozen scallop and US$20/kg for live scallop in 2019 [10]. Many species also have an esthetic value, providing ornamental objects such as pearls and decorative shells. Worldwide production of unworked pearls reached US$412 million and that of worked pearls was US$ 787 million in 2004 [11], promoting the fishing industry in coastal zones [11, 12]. In Europe, bivalve production in aquaculture systems is reported to be about 598 thousand tons per year, and global production exceeds 15 million tons, translating to about 20.6 billion dollars per year worldwide [13]. Bivalves' application for the human wellness are many, such as the ornamental and food resources, including the extraction of bioactive molecules with medical use, due to its antiviral [14] and antibacterial (e.g., the lectin MytiLec-1 extracted from *Mytilus galloprovincialis* and with the ability to inhibit the growth of Gram-positive and Gram-negative bacteria) [15] properties. They are also able to cause the apoptosis of cancer cells [16–18] and have the potential to the development of novel food products, due to antioxidant extracts from the bivalves shell [19–21]. Considering the important ecological and economic value of these organisms, the growing deterioration of bivalves' communities should also be a major concern to all and thus to the achievement of the goals of Agenda 2030.

4. Anthropogenic pressures

Anthropogenic activities, particularly the industrial and urban development, agricultural practices, tourism, and recreational aquatic activities [22–24] contribute to the aquatic ecosystems' pollution, as well as the exponential increase of the human population. Consequently, the food need leads to organisms' overexploitation. These pressures, pairwise the introduction of alien species, accidental or per esthetic and economic reasons, are leading to the habitats' degradation, affecting the aquatic ecosystems [25, 26] and interact by complex and unpredictable ways, with dangerous consequences to the systems, affecting their stability and resilience [27], resulting in bivalves communities' decline. Several researchers [24, 28–30] have dedicated their studies to understand the effects of environmental contaminants on aquatic ecosystems and consequently on the species.

Despite the pressures resulting directly from human activities, aquatic ecosystems are also very susceptible to climatic changes. Nowadays, global changes comprise a main concern to scientific, political, and social communities. IPCC [31] predicts a temperature increase of the Earth's surface of about 4°C until the end of twenty-first century, with dangerous impacts to the aquatic systems and organisms, being particularly aggravated by bivalves' species [32].

The main reasons and effects of the above-mentioned stressors are detailed in the further subsections, considering its impacts on the aquatic systems and particularly to the bivalves' populations. The selection of stressors was based on its increase and harmful effects, and comprises the: i) chemical contamination (including pharmaceuticals, pesticides, and metals) and microplastics, highly used in daily activities by human beings (e.g., pesticides used worldwide increased 1 million tons in the last 30 years) [33]; ii) overexploitation of resources, resulting from the increasing of the food requirements [34]; and iii) the climatic changes (**Figure 1**) [31, 35].

4.1 Contaminants

Many effects have been reported at the individual level (e.g., inhibition of the growth, reproduction, behavior, and survival) [22, 24, 30, 36], being usually associated with causes at a molecular level, such as changes in detoxication, neurological, immunological, metabolic and antioxidant processes, biomolecules, and DNA [29, 37–47].

4.1.1 Pharmaceuticals

Chemicals designed to act on specific molecular and metabolic pathways of animals and humans, nevertheless, may affect non-target species [29]. These contaminants enter on the aquatic systems mostly by industrial effluents, livestock wastes, human excretions, hospital wastes, and incorrect discarding of pharmaceuticals, incomplete drugs removal of many medicines, by wastewater treatment plants, and release of contaminated effluents and sludge containing drugs [48, 49].

4.1.2 Pesticides

Chemical formulations, mostly from agricultural use for pest control (including insects, weeds, parasites, fungi and rodents) [50], may affect the organisms directly, by their toxic effects, or indirectly, by the elimination of other species [51].

Figure 1.
Bivalves ecological roles and the impacts and pressures of human activities.

4.1.3 Metals

Metals can be from industrial and mining activity, metal plating, rainwater runoff, and sewage [52–54], or released by natural sources [55]. These compounds can be classified into two categories—essential metals and non-essential metals. Essential metals have a biological role in the organisms (e.g., copper, iron, magnesium and zinc) [56, 57], but still they may be toxic at high concentrations. Non-essential metals have no role in organic processes (e.g., mercury, lead, cadmium, and arsenic) [58, 59], being toxic even at very low concentrations [60].

4.1.4 Microplastics

These emerging contaminants are a worldwide concern among political and scientific communities, but also to the general public [61, 62]. These particulates may be from the degradation of bigger plastic elements or plastic initially produced with a little dimension for commercial application [8]. They are easily ingested by organisms from various trophic levels [63–66] and have the ability to accumulate in bivalves [28], with potential consequences to human health. Fang et al. [67] reported an ingestion of 2565 microplastics items by volunteers, after consumption of 225 g of mussels, an exposure classified as considerable [68, 69]. Moreover, studies estimate an annual ingestion of microplastics per person, as consequence of the shellfish consumption, of about 87 to 11,970 items [70]. The interest in microplastics studies is increasing, and some researchers have observed dangerous effects on bivalves, as a consequence of the exposure to these particles. Browne et al. [71] demonstrated microplastics translocation to hemolymph and to hemocytes. Structural and morphological changes on tissues and effects to the gills and digestive gland were shown by Alnajar et al. [28]. Other authors also stated reactive oxygen species production, oxidative stress generation, and DNA chain breaks [72, 73].

The above-mentioned contaminants have dangerous effects on bivalves' species and the remaining aquatic communities, and on water quality and consequently on human beings. Considering the 2.2 billion of people without safe access to drinking water and the resources provided by aquatic systems, the preservation of these ecosystems is essential. Some policy measures, such as fines, have been implemented to control contaminant discharges to aquatic systems; however, this problem remains. Therefore, aggravation of fines for those who do not comply, more intensive supervision, and more restrictive laws are necessary to ensure availability and sustainable management of water, and also to achieve a sustainable use of the oceans and the marine resources. Moreover, measures like attribution of economic benefits for the safe disposal of pollutants could be implemented, helping to restore an appropriate water quality. Another measure to reduce pollution is the application of natural fertilizers in agriculture fields, which are healthier for soils, cultivated products (also increasing their quality), human health, and the surrounding aquatic systems [74]. Improved irrigation techniques, like those already used in some countries where water resources availability is a problem, can help to reduce agricultural water use. These are some of the measures to achieve some of the targets of the SDGs 6 and 14 of Agenda 2030.

4.2 Overexploitation of resources

During the past century, the development and usage of resources to improve living standards proceeded rapidly, in some cases without concern for the sustainability of the systems and resources. The management and implementation of mitigating measures in threatened aquatic ecosystems were also conducted to restore ecological health, mainly in those surrounded by agriculture fields with intensive practice. The overexploitation of agriculture fields hasted to a depletion of land resources, which poses a new challenge to develop new food sources focused in marine resources [75]. Contrary to land systems, in the ocean the access to resources outside of national jurisdiction is practically unregulated. Furthermore, in territorial waters and exclusive economic zones, where there is regulation, it is often ineffective. Several consequences are frequently associated with non-regulation, such as overfishing, blocking of maritime shipping ways, coastal habitats destruction, introduction of dangerous alien species, and occasional and pollution above the resource assimilation ability [76].

Some researchers have reported population changes associated with overexploitation, namely the decrease of demographic, temporal, and spatial diversity [77], changes in population age and size, decrease of genetic diversity, and changes in growth rate and reproduction [78]. Moreover, variations in trophic structure as consequence of overexploitation are reported [79–82], as well as, biomass decline [83].

Among the many methods of overexploitation of resources, many research studies have been focused on fisheries using equipment that damage benthic habitats and pose risks to the administration, balanced use, and preservation of live resources [84–87]. Hand collection of mollusks (mussels, clams and oysters) is among the most susceptible of artisanal fisheries, due to its distribution and economic value in the coastal zones [88–90].

To achieve SDG 14, some measures (e.g., attribution of economic benefits to the sustainable use of marine resources, prohibition of subsidies that contribute to the overfishing, determination of fishing quotes) [91] should be applied to support sustainable fishing that ensures the continuation of the species and protect aquatic communities. However, fishing is the livelihood of many people, who often lack access to basic information, and therefore knowledge-sharing programs highlighting the importance of the ecosystems and resources preservation are badly needed.

4.3 Climatic changes

Nowadays, climatic changes are the main concern among political and scientific communities, with SDG 13 focusing on urgent action to combat climate change and its impacts. It is recognized that climatic changes affect all life forms, and marine ecosystems are not an exception. So several studies have showed the climatic change impacts on these systems [92, 93], namely warming and acidification of the oceans [94], changes in the nutrient dynamics, increase of CO_2 levels, and alterations in wind and flow patterns [95]. Consequently, the survival, growth, reproduction, and distribution of species are affected, and the impacts also may be observed at higher levels such as population, community, and ecosystem [96]. Some studies have linked climatic changes with dangerous biological effects on bivalve species, such as shell formation [94], abundance [97], development [98], larval stage [99], recruitment [100–102], and spawning [103].

Ecosystem monitoring is fundamental to understand the long-term evolution of ecosystems, allowing the detection of changes in biotic and abiotic parameters and intervention to restore ecosystem balance. Some efforts have been done by scientific research groups to monitoring these ecosystems, such as determining and quantification of metals [104–106], organic pollutants [107–109], and microplastics [110, 111]. The ocean acidification impact [112, 113], the development of technology [114], or even creation of marine sanctuaries [115] are other contributions to the monitoring programs. However, these programs are poorly funded by governments and are mainly dependent on the scientific research efforts, being crucial a more effective and higher financial support.

5. Conclusion

Anthropogenic pressures on aquatic systems are increasingly intense and need better understanding of their impacts on ecosystems and communities. This review highlights the dangerous effects of pressures from human activities, with a focus on contaminant inputs, resource overexploitation, and climatic changes. Moreover, this work intends to be an alert to the importance of aquatic ecosystems' recovery and preservation, and thus water quality, with a focus on the ecological role of bivalve species. This work also calls attention to the decline of these communities and the impacts on the ecosystem structure and functioning, not only in marine systems but also in freshwater environments, where bivalves may have a great potential for water cleaning and purification. Human beings have a key role in the preservation and conservation of aquatic systems, species, and water quality through the application of mitigation measurements and strategies. It is crucial to act to achieve the goals and targets proposed in Agenda 2030 for a better and a more sustainable future. The future depends on us!

Acknowledgements

This study was supported by Fundação para a Ciência e a Tecnologia (FCT) through the strategic projects LA/P/0069/2020 granted to the Associate Laboratory ARNET, UIDB/04292/2020 granted to the Marine and Environmental Sciences Centre (MARE) and UIDP/50017/2020 + UIDB/50017/2020+ LA/P/0094/2020 granted to CESAM. A. F. Mesquita thanks FCT for the financial support provided through doctoral grant SFRH/BD/139831/2018, co-funded by the Human Potential Operational Programme (National Strategic Reference Framework 2007–2013), European Social Fund (EU), and the program POPH/FSE. A. M. M. Gonçalves acknowledges the University of Coimbra for contract IT057-18-7253.

Author contributions

Conceived and designed the idea, A.F.M. and A.M.M.G.; organization of the team, A.M.M.G.; writing and bibliographic research, A.F.M.; supervision and manuscript revision, F.J.M.G. and A.M.M.G. All authors have read and agreed to the published version of the manuscript.

Author details

Andreia Filipa Mesquita[1], Fernando José Mendes Gonçalves[1] and
Ana Marta Mendes Gonçalves[1,2]*

1 Department of Biology and CESAM, University of Aveiro, Aveiro, Portugal

2 University of Coimbra, MARE - Marine and Environmental Sciences Centre / ARNET - Aquatic Research Network, Department of Life Sciences, Calçada Martim de Freitas, Coimbra, Portugal

*Address all correspondence to: amgoncalves@uc.pt

References

[1] Schwan G. Sustainable development goals. Gaia. 2019;**28**(2):73. DOI: 10.14512/gaia.28.2.1

[2] NOAA. National Ocean Service Website. Where is all of the Earth's Water? 2020. Available from: https://oceanservice.noaa.gov/facts/wherewater.html

[3] Halpern BS, Longo C, Hardy D, McLeod KL, Samhouri JF, Katona SK, et al. An index to assess the health and benefits of the global ocean. Nature. 2012;**488**(7413):615-620. DOI: 10.1038/nature11397

[4] Harris P, Tuhumwire J. Part II - Chapter I: Introduction - Planet, oceans and life. In: The First Global Integrated Marine Assessment World Ocean Assessment I. United Nations. 2016. Available from: https://www.un.org/depts/los/global_reporting/WOA_RPROC/WOACompilation.pdf

[5] Vaughn CC, Hoellein TJ. Bivalve impacts in freshwater and marine ecosystems. Annual Review of Ecology, Evolution, and Systematics. 2018;**49**:183-208. DOI: 10.1146/ANNUREV-ECOLSYS-110617-062703. [accessed 2021 Oct 20]. Available from: https://www.annualreviews.org/doi/abs/10.1146/annurev-ecolsys-110617-062703

[6] Sicuro B, Castelar B, Mugetti D, Pastorino P, Chiarandon A, Menconi V, et al. Bioremediation with freshwater bivalves: A sustainable approach to reducing the environmental impact of inland trout farms. Journal of Environmental Management. 2020;**276**:111327. DOI: 10.1016/j.jenvman.2020.111327

[7] Waser AM, Deuzeman S, Kangeri AKW, van Winden E, Postma J, de Boer P, et al. Impact on bird fauna of a non-native oyster expanding into blue mussel beds in the Dutch Wadden Sea. Biological Conservation. 2016;**202**:39-49. DOI: 10.1016/j.biocon.2016.08.007

[8] Zhang F, Man YB, Mo WY, Man KY, Wong MH. Direct and indirect effects of microplastics on bivalves, with a focus on edible species: A mini-review. Critical Reviews in Environmental Science and Technology. 2020;**50**(20):2109-2143. DOI: 10.1080/10643389.2019.1700752. [accessed 2020 Oct 13]. Available from: https://www.tandfonline.com/doi/abs/10.1080/10643389.2019.1700752

[9] FAO. Bivalve Trade is Growing. 2018. Available from: http://www.fao.org/in-action/globefish/market-reports/resource-detail/en/c/1107016/ [Accessed: May 24, 2021]

[10] FAO. World Bivalve Market Heavily Impacted by COVID-19. 2021. Available from: http://www.fao.org/in-action/globefish/market-reports/resource-detail/en/c/1296658/ [Accessed: May 24, 2021]

[11] Tisdell CA, Poirine B. Economics of Pearl Farming - HAL-SHS - Sciences de l'Homme et de la Société. 2008. Available from: https://halshs.archives-ouvertes.fr/halshs-01141429 [Accessed: October 20, 2020]

[12] Gopalakrishnan S, Vijayavel K. Nutritional composition of three estuarine bivalve mussels, *Perna viridis*, *Donax cuneatus* and *Meretrix meretrix*. International Journal of Food Sciences and Nutrition. 2009;**60**(6):458-463. DOI: 10.1080/09637480701830412. [accessed 2020 Oct 20]. Available from: http://www.tandfonline.com/doi/full/10.1080/09637480701830412

[13] Smaal AC, Ferreira JG, Grant J, Petersen JK, Strand O. Goods and Services of Marine Bivalves. Cham, Switzerland: Springer Open; 2019

[14] Lin-Rui C, Qing L, Zhen-Xing S, Yu-Chao Y, Lin S. Antioxidant and antibacterial activities of polysaccharides from Chinese surf clam (*Mactra chinensis*). Food Science. 2012;**33**(7):101-104. [accessed 2023 Feb 18]. Available from: https://www.scopus.com/record/display.uri?eid=2-s2.0-84969330898&origin=inward&txGid=3606bbeed410890a7afd7ae8a5dce601

[15] Hasan I, Gerdol M, Fujii Y, Rajia S, Koide Y, Yamamoto D, et al. cDNA and gene structure of MytiLec-1, a bacteriostatic R-type lectin from the Mediterranean mussel (*Mytilus galloprovincialis*). Marine Drugs. 2016;**14**(5):92. DOI: 10.3390/MD14050092. [accessed 2023 Feb 18]. Available from: https://pubmed.ncbi.nlm.nih.gov/27187419/

[16] Chernikov O, Kuzmich A, Chikalovets I, Molchanova V, Hua KF. Lectin CGL from the sea mussel *Crenomytilus grayanus* induces Burkitt's lymphoma cells death via interaction with surface glycan. International Journal of Biological Macromolecules. 2017;**104**(Pt A):508-514. DOI: 10.1016/J.IJBIOMAC.2017.06.074. [accessed 2023 Feb 18]. Available from: https://pubmed.ncbi.nlm.nih.gov/28636877/

[17] Hasan I, Sugawara S, Fujii Y, Koide Y, Terada D, Iimura N, et al. MytiLec, a mussel R-type lectin, interacts with surface glycan Gb3 on Burkitt's lymphoma cells to trigger apoptosis through multiple pathways. Marine Drugs. 2015;**13**(12):7377-7389. DOI: 10.3390/MD13127071. [accessed 2023 Feb 18]. Available from: https://pubmed.ncbi.nlm.nih.gov/26694420/

[18] Liao JH, Chien CTH, Wu HY, Huang KF, Wang I, Ho MR, et al. A multivalent marine lectin from *Crenomytilus grayanus* possesses anti-cancer activity through recognizing Globotriose Gb3. Journal of the American Chemical Society. 2016;**138**(14):4787-4795. DOI: 10.1021/JACS.6B00111/SUPPL_FILE/JA6B00111_SI_001.PDF. [accessed 2023 Feb 18]. Available from: https://pubs.acs.org/doi/abs/10.1021/jacs.6b00111

[19] Chai TT, Law YC, Wong FC, Kim SK. Enzyme-assisted discovery of antioxidant peptides from edible marine invertebrates: A review. Marine Drugs. 2017;**15**(2):42. DOI: 10.3390/MD15020042. [accessed 2023 Feb 18]. Available from: https://pubmed.ncbi.nlm.nih.gov/28212329/

[20] Odeleye T, Li Y, White WL, Nie S, Chen S, Wang J, et al. The antioxidant potential of the New Zealand surf clams. Food Chemistry. 2016;**204**:141-149. DOI: 10.1016/J.FOODCHEM.2016.02.120

[21] Wu S, Huang X. Preparation and antioxidant activities of oligosaccharides from *Crassostrea gigas*. Food Chemistry. 2017;**216**:243-246. DOI: 10.1016/J.FOODCHEM.2016.08.043

[22] Gonçalves AMM, Mesquita AF, Verdelhos T, Coutinho JAP, Marques JC, Gonçalves F. Fatty acids' profiles as indicators of stress induced by of a common herbicide on two marine bivalves species: *Cerastoderma edule* (Linnaeus, 1758) and *Scrobicularia plana* (da costa, 1778). Ecological Indicators. 2016;**63**:209-218. DOI: 10.1016/J.ECOLIND.2015.12.006. [accessed 2019 Feb 20]. Available from: https://www.sciencedirect.com/science/article/pii/S1470160X15007086

[23] Maria VL, Bebianno MJ. Antioxidant and lipid peroxidation responses

in *Mytilus galloprovincialis* exposed to mixtures of benzo(a)pyrene and copper. Comparative Biochemistry and Physiology Part C: Toxicology & Pharmacology. 2011;**154**(1):56-63. DOI: 10.1016/j.cbpc.2011.02.004

[24] Mesquita AF, Gonçalves F, Verdelhos T, Marques JC, Gonçalves AMM. Fatty acids profiles modifications in the bivalves *Cerastoderma edule* and *Scrobicularia plana* in response to copper sulphate. Ecological Indicators. 2018;**85**(October 2017):318-328. DOI: 10.1016/j.ecolind.2017.10.007

[25] Halpern BS, McLeod KL, Rosenberg AA, Crowder LB. Managing for cumulative impacts in ecosystem-based management through ocean zoning. Ocean and Coastal Management. 2008;**51**(3):203-211. DOI: 10.1016/j.ocecoaman.2007.08.002

[26] Schwarzenbach RP, Egli T, Hofstetter TB, von Gunten U, Wehrli B. Global water pollution and human health. Annual Review of Environment and Resources. 2010;**35**(1):109-136. DOI: 10.1146/annurev-environ-100809-125342. [accessed 2019 Feb 22]. Available from: http://www.annualreviews.org/doi/10.1146/annurev-environ-100809-125342

[27] Beaugrand G, Edwards M, Legendre L. Marine biodiversity, ecosystem functioning, and carbon cycles. Proceedings of the National Academy of Sciences of the United States of America. 2010;**107**(22):10120-10124. DOI: 10.1073/pnas.0913855107. [accessed 2020 Oct 22]. Available from: https://www.pnas.org/content/107/22/10120

[28] Alnajar N, Jha AN, Turner A. Impacts of microplastic fibres on the marine mussel, *Mytilus galloprovinciallis*. Chemosphere. 2021;**262**:128290. DOI: 10.1016/j.chemosphere.2020.128290

[29] Munari M, Matozzo V, Chemello G, Riedl V, Pastore P, Badocco D, et al. Seawater acidification and emerging contaminants: A dangerous marriage for haemocytes of marine bivalves. Environmental Research. 2019;**175**:11-21. DOI: 10.1016/j.envres.2019.04.032

[30] Solé M, Shaw JP, Frickers PE, Readman JW, Hutchinson TH. Effects on feeding rate and biomarker responses of marine mussels experimentally exposed to propranolol and acetaminophen. Analytical and Bioanalytical Chemistry. 2010;**396**:649-656. [accessed 2020 Oct 25]. Available from: https://link.springer.com/article/10.1007/s00216-009-3182-1

[31] IPCC. Climate change 2014: Synthesis report. In: Contribution of Working Groups I, IIand III to the Fifth Assessment Report of the Intergovernmental Panel on ClimateChange. Geneva: Cambridge University Press; 2014

[32] IPCC. Global warming of 1.5° C. In: An IPCC Special Report on the impacts of global warming of 1.5°C. Intergovernamental Panel on Climate Change. Paris; 2018

[33] FAO. Pesticides Use. 2020. Available from: https://www.fao.org/faostat/en/#data/RP/visualize [Accessed: January 16, 2023]

[34] United Nations. World Population Prospects. 2022. Available from: https://population.un.org/wpp/ [Accessed: January 24, 2023]

[35] Tollefson J. How hot will Earth get by 2100? Nature. 2020;**580**(7804):443-445. DOI: 10.1038/D41586-020-01125-X

[36] Neves M, Castro BB, Vidal T, Vieira R, Marques JC, Coutinho JAP, et al. Biochemical and populational responses of an aquatic bioindicator species, *Daphnia longispina*, to a commercial

formulation of a herbicide (Primextra® gold TZ) and its active ingredient (S-metolachlor). Ecological Indicators. 2015;**53**:220-230. DOI: 10.1016/J.ECOLIND.2015.01.031. [accessed 2019 Feb 20]. Available from: https://www.sciencedirect.com/science/article/pii/S1470160X15000552

[37] Brahma N, Gupta A. Acute toxicity of lead in fresh water bivalves *Lamellidens jenkinsianus obesa and Parreysia (Parreysia) corrugata* with evaluation of sublethal effects on acetylcholinesterase and catalase activity, lipid peroxidation, and behavior. Ecotoxicology and Environmental Safety. 2020;**189**:109939. DOI: 10.1016/j.ecoenv.2019.109939

[38] Carregosa V, Velez C, Soares AMVM, Figueira E, Freitas R. Physiological and biochemical responses of three Veneridae clams exposed to salinity changes. Comparative Biochemistry and Physiology - B Biochemistry and Molecular Biology. 2014;**177-178**:1-9. DOI: 10.1016/j.cbpb.2014.08.001. [accessed 2017 Mar 13]. Available from: http://www.ncbi.nlm.nih.gov/pubmed/25132624

[39] Cossu C, Doyotte A, Babut M, Exinger A, Vasseur P. Antioxidant biomarkers in freshwater bivalves, *Unio tumidus*, in response to different contamination profiles of aquatic sediments. Ecotoxicology and Environmental Safety. 2000;**45**(2):106-121. DOI: 10.1006/eesa.1999.1842

[40] Fabbri E, Franzellitti S. Human pharmaceuticals in the marine environment: Focus on exposure and biological effects in animal species. Environmental Toxicology and Chemistry. 2016;**35**(4):799-812. DOI: 10.1002/etc.3131. [accessed 2020 Oct 20]. Available from: https://pubmed.ncbi.nlm.nih.gov/26111460/

[41] Gonzalez-Rey M, Bebianno MJ. Effects of non-steroidal anti-inflammatory drug (NSAID) diclofenac exposure in mussel *Mytilus galloprovincialis*. Aquatic Toxicology. 2014;**148**:818-831. DOI: 10.1016/j.aquatox.2014.01.011

[42] Gutiérrez IB, Mesquita AF, Gonçalves FJM, Marques JC, Gonçalves AMM. Biomarkers' responses of the benthic clam *Scrobicularia plana* to the main active ingredients (S-metolachlor and Terbuthylazine) of a common herbicide. Ecological Indicators. 2019;**96**:611-619. DOI: 10.1016/j.ecolind.2018.09.044

[43] Martin-Diaz L, Franzellitti S, Buratti S, Valbonesi P, Capuzzo A, Fabbri E. Effects of environmental concentrations of the antiepilectic drug carbamazepine on biomarkers and cAMP-mediated cell signaling in the mussel *Mytilus galloprovincialis*. Aquatic Toxicology. 2009;**94**(3):177-185. DOI: 10.1016/j.aquatox.2009.06.015

[44] Mesquita AF, Marques SM, Marques JC, Gonçalves FJM, Gonçalves AMM. Copper sulphate impact on the antioxidant defence system of the marine bivalves *Cerastoderma edule* and *Scrobicularia plana*. Scientific Reports. 2019;**9**(1):1-11. DOI: 10.1038/s41598-019-52925-9

[45] Mezzelani M, Gorbi S, Regoli F. Pharmaceuticals in the aquatic environments: Evidence of emerged threat and future challenges for marine organisms. Marine Environmental Research. 2018;**140**:41-60. DOI: 10.1016/j.marenvres.2018.05.001

[46] Patil PB, Mahajan AY. Acute toxicity of PROFENOFOS for freshwater mussel bivalve *Lammellidens corrianus*. Advances in Pharmacology and Toxicology. 2012;**13**(3):41-45. [accessed 2020 Nov 14]. Available from: https://eds.b.ebscohost.com/eds/pdfviewer/pdfviewer?vid=0&sid=1c53f985-4f69-

4e29-bff5-5733306b3850%40pdc-v-sessmgr05

[47] Sheir SK. Histological responses of the freshwater clam, *Caelatura nilotica*. Different environmental degrees of mixed pollution. Egyptian Academic Journal of Biological Sciences. 2020;**12**(2):49-66

[48] Monteiro SC, Boxall ABA. Occurrence and fate of human Pharmaceuticals in the Environment. In: Reviews of Environmental Contamination and Toxicology. Vol. 202. New York, NY: Springer; 2010. pp. 53-154. [accessed 2020 Oct 20]. Available from: https://link.springer.com/chapter/10.1007/978-1-4419-1157-5_2

[49] Rzymski P, Drewek A, Klimaszyk P. Pharmaceutical pollution of aquatic environment: An emerging and enormous challenge. Limnology Review. 2017;**17**(2):97-107. DOI: 10.1515/limre-2017-0010. [accessed 2020 Oct 22]. Available from: https://content.sciendo.com/view/journals/limre/17/2/article-p97.xml

[50] World Health Organization. Guidelines for Drinking-Water Quality. Geneva, Switzerland: World Health Organization; 2011

[51] Renault T. Immunotoxicological effects of environmental contaminants on marine bivalves. Fish & Shellfish Immunology. 2015;**46**(1):88-93. DOI: 10.1016/j.fsi.2015.04.011

[52] El-Rayis OA, Abouldahab O, Halim Y, Riley JP. Levels of trace metals in some food chain organisms from el-mex bay, west of alexandria, egypt. In: Proceedings of the 7th International Conference: Environment Protection is a Must. 1997. pp. 20-22. Available from: https://www.scopus.com/record/display.uri?eid=2-s2.0-84865766244&origin=inward. [Accessed: February 21, 2019]

[53] Mance G. Introduction. In: Pollution Threat of Heavy Metals in Aquatic Environments. Dordrecht: Springer Netherlands; 1987. pp. 1-8. [accessed 2019 Feb 21]. http://www.springerlink.com/index/10.1007/978-94-009-3421-4_1

[54] Soualili D, Dubois P, Gosselin P, Pernet P, Guillou M. Assessment of seawater pollution by heavy metals in the neighbourhood of Algiers: Use of the sea urchin, *Paracentrotus lividus*, as a bioindicator. ICES Journal of Marine Science. 2008;**65**(2):132-139. DOI: 10.1093/icesjms/fsm183. [accessed 2019 Feb 21]. Available from: https://academic.oup.com/icesjms/article-lookup/doi/10.1093/icesjms/fsm183

[55] Nekoubin H, Gharedaashi E, Hatefi S, Sudagar M, Shahriari R, Asgharimoghadam A. Determination of LC 50 of copper Sulfate and Lead (II) nitrate and Behavioral responses of grass carp (*Ctenopharyngodon idella*). Walailak Journal of Science and Technology (WJST). 2012;**9**(Ii):333-340

[56] Simkiss K. Cellular discrimination processes in metal accumulating cells. The Journal of Experimental Biology. 1981;**94**(1):317-327

[57] Williams RJP, Coombs TL, Tinker PB. Physico-chemical aspects of inorganic element transfer through membranes [and discussion]. Philosophical Trans actions of the Royal Society B Biology Science. 1981;**294**(1071):57-74. DOI: 10.1098/rstb.1981.0089. [accessed 2019 Feb 21]. Available from: http://rstb.royalsocietypublishing.org/cgi/doi/10.1098/rstb.1981.0089

[58] Dallinger R. Metabolism an toxicity of metals: Metallothionines and metal

elimination. In: Cajaraville MP, editor. Cell Biology in Environment Toxicology. Bilbao: Universidade del Pais Vasco; 1995. pp. 171-190

[59] Suzuki KT, Suzuki T. An introduction to clinical aspects of toxicity. In: Chang W, editor. Toxicology of Metals. Boca Raton: LCRC Lewis Publishers; 1996. pp. 333-335

[60] Bae JH, Lim SY. Heavy metals and biochemical composition of four sea bream species (*Acanthopagrus schlegelii* Bleeker, *Pagrus major* Temminck & Schlegel, *Oplegnathus fasciatus* Krøyer and *Girella punctata* Gray). Philippines Agricultureal Science. 2012;**95**(2):185-191

[61] Lusher A, Hollman P, Mendonza-Hill J. Microplastics in fisheries and aquaculture status of knowledge on their occurrence and implications for aquatic organisms and food safety. Food Agriculture Organic United Nations. 2017;**615**:1-4

[62] Seltenrich N. New link in the food chain? Marine plastic pollution and seafood safety. Environmental Health Perspectives. 2015;**123**(2):A34–A41. DOI: 10.1289/ehp.123-A34. [accessed 2020 Oct 20]. Available from: https://www.ncbi.nlm.nih.gov/pmc/articles/PMC4314237/

[63] Fang C, Zheng R, Zhang Y, Hong F, Mu J, Chen M, et al. Microplastic contamination in benthic organisms from the Arctic and sub-Arctic regions. Chemosphere. 2018;**209**:298-306. DOI: 10.1016/j.chemosphere.2018.06.101

[64] Li HX, Ma LS, Lin L, Ni ZX, Xu XR, Shi HH, et al. Microplastics in oysters *Saccostrea cucullata* along the Pearl River estuary, China, Environment Pollution. 2018;**236**:619-625. DOI: 10.1016/j.envpol.2018.01.083. [accessed 2020 Oct 20]. Available from: https://pubmed.ncbi.nlm.nih.gov/29433102/

[65] Morgana S, Ghigliotti L, Estévez-Calvar N, Stifanese R, Wieckzorek A, Doyle T, et al. Microplastics in the Arctic: A case study with sub-surface water and fish samples off Northeast Greenland. Environmental Pollution. 2018;**242**:1078-1086. DOI: 10.1016/j.envpol.2018.08.001

[66] Rezania S, Park J, Md Din MF, Mat Taib S, Talaiekhozani A, Kumar Yadav K, et al. Microplastics pollution in different aquatic environments and biota: A review of recent studies. Marine Pollution Bulletin. 2018;**133**:191-208. DOI: 10.1016/j.marpolbul.2018.05.022

[67] Fang C, Zheng R, Chen H, Hong F, Lin L, Lin H, et al. Comparison of microplastic contamination in fish and bivalves from two major cities in Fujian province, China and the implications for human health. Aquaculture. 2019;**512**:734322. DOI: 10.1016/j.aquaculture.2019.734322

[68] Barboza LGA, Dick Vethaak A, Lavorante BRBO, Lundebye AK, Guilhermino L. Marine microplastic debris: An emerging issue for food security, food safety and human health. Marine Pollution Bulletin. 2018;**133**:336-348. DOI: 10.1016/j.marpolbul.2018.05.047

[69] EFSA. Presence of microplastics and nanoplastics in food, with particular focus on seafood. EFSA Journal. 2016;**14**(6):4501. DOI: 10.2903/j.efsa.2016.4501. [accessed 2020 Oct 20]. Available from: http://doi.wiley.com/10.2903/j.efsa.2016.4501

[70] Cho Y, Shim WJ, Jang M, Han GM, Hong SH. Abundance and characteristics of microplastics in market bivalves from South Korea. Environmental Pollution.

2019;**245**:1107-1116. DOI: 10.1016/j.envpol.2018.11.091

[71] Browne MA, Dissanayake A, Galloway TS, Lowe DM, Thompson RC. Ingested microscopic plastic translocates to the circulatory system of the mussel, *Mytilus edulis* (L.). Environmental Science & Technology. 2008;**42**(13):5026-5031. DOI: 10.1021/es800249a. [accessed 2020 Oct 22]. Available from: https://pubs.acs.org/sharingguidelines

[72] Avio CG, Gorbi S, Milan M, Benedetti M, Fattorini D, D'Errico G, et al. Pollutants bioavailability and toxicological risk from microplastics to marine mussels. Environmental Pollution. 2015;**198**:211-222. DOI: 10.1016/j.envpol.2014.12.021

[73] Ribeiro F, Garcia AR, Pereira BP, Fonseca M, Mestre NC, Fonseca TG, et al. Microplastics effects in *Scrobicularia plana*. Marine Pollution Bulletin. 2017;**122**(1-2):379-391. DOI: 10.1016/j.marpolbul.2017.06.078. [accessed 2020 Oct 20]. Available from: https://pubmed.ncbi.nlm.nih.gov/28684108/

[74] de Melo PC, Collela CF, Sousa T, Pacheco D, Cotas J, Gonçalves AMM, et al. Seaweed-based products and mushroom β-glucan as tomato plant immunological inducers. Vaccine. 2020;**8**(3):1-13. DOI: 10.3390/vaccines8030524

[75] Zhang S, Zhou H. Evaluation of the influence of various factors on project performance under the EPC general contracting mode in the development of marine resources on JSTOR. Journal of Coastal Research. 2019;**93**:762-767. [accessed 2020 Oct 22]. Available from: https://www.jstor.org/stable/26853350?seq=1

[76] Berkes F, Mahon R, Mcconney P, Pollnac R, Pomeroy R. Key concepts in fisheries management. In: Managing Small-Scale Fisheries Alternative Directions and Methods Contents. Ottawa, Canada: International Development Research Centre; 2010

[77] Perry RI, Cury P, Brander K, Jennings S, Möllmann C, Planque B. Sensitivity of marine systems to climate and fishing: Concepts, issues and management responses. Journal of Marine Systems. 2010;**79**(3-4):427-435. DOI: 10.1016/j.jmarsys.2008.12.017

[78] Goñi R. Ecosystem effects of marine fisheries: An overview. Ocean and Coastal Management. 1998;**40**(1):37-64. DOI: 10.1016/S0964-5691(98)00037-4

[79] Blanchard JL, Dulvy NK, Jennings S, Ellis JR, Pinnegar JK, Tidd A, et al. Do climate and fishing influence size-based indicators of Celtic Sea fish community structure? ICES Journal of Marine Science. 2005;**62**(3):405-411. DOI: 10.1016/j.icesjms.2005.01.006. [accessed 2020 Oct 22]. Available from: https://academic.oup.com/icesjms/article/62/3/405/661410

[80] Gascuel D, Coll M, Fox C, Guénette S, Guitton J, Kenny A, et al. Fishing impact and environmental status in European seas: A diagnosis from stock assessments and ecosystem indicators. Fish and Fisheries. 2016;**17**(1):31-55. DOI: 10.1111/faf.12090. [accessed 2020 Oct 22]. Available from: http://doi.wiley.com/10.1111/faf.12090

[81] Pinnegar JK, Jennings S, O'Brien CM, Polunin NVC. Long-term changes in the trophic level of the Celtic Sea fish community and fish market price distribution. Journal of Applied Ecology. 2002;**39**(3):377-390. DOI: 10.1046/j.1365-2664.2002.00723.x

[82] Shephard S, Fung T, Houle JE, Farnsworth KD, Reid DG, Rossberg AG. Size-selective fishing drives species composition in the Celtic Sea. ICES Journal of Marine Science. 2012;**69**(2):223-234. DOI: 10.1093/icesjms/fsr200. [accessed 2020 Oct 22]. Available from: http://www.fishbase.org

[83] Hernvann PY, Gascuel D. Exploring the impacts of fishing and environment on the Celtic Sea ecosystem since 1950. Fisheries Research. 2020;**225**:105472. DOI: 10.1016/j.fishres.2019.105472

[84] Dayton PK, Thrush SF, Tundi Agardy M, Hofman RJ. Environmental effects of marine fishing. Aquatic Conservation: Marine and Freshwater Ecosystems. 1995;**5**:205-232

[85] Jennings S, Kaiser MJ. The effects of fishing on marine ecosystems. Advances in Marine Biology. 1998;**34**(34):201-212. DOI: 10.1016/s0065-2881(08)60212-6

[86] Kappel CV. Losing pieces of the puzzle: Threats to marine, estuarine, and diadromous species. Frontiers in Ecology and the Environment. 2005;**3**(5):275-282. DOI: 10.1890/1540-9295(2005)003[0275:LPOTPT]2.0.CO;2. [accessed 2020 Oct 20]. Available from: https://esajournals.onlinelibrary.wiley.com/doi/full/10.1890/1540-9295%282005%29003%5B0275%3ALPOTPT%5D2.0.CO%3B2

[87] Shester GG, Micheli F. Conservation challenges for small-scale fisheries: Bycatch and habitat impacts of traps and gillnets. Biological Conservation. 2011;**144**(5):1673-1681. DOI: 10.1016/j.biocon.2011.02.023

[88] Berkes F, Mahon R, McConney P, Pollnac R, Pomeroy R. Managing Small-Scale Fisheries: Alternative Directions and Methods | IDRC - International Development Research Centre. 2001. Available from: https://www.idrc.ca/en/book/managing-small-scale-fisheries-alternative-directions-and-methods [Accessed: October 22, 2020]

[89] Berthou P, Poutiers J, Goulletquer P, Dao J. Shelled molluscs. In: FISHERIES AND AQUACULTURE, from Encyclopedia of Life Support Systems (EOLSS), Developed under the Auspices of the UNESCO. Oxford, UK: Eolss Publishers; 2009. Available from: http://www.eolss.net

[90] Toral-Granda V, Lovatelli A, Vasconcellos M. Sea cucumbers - a global review of fisheries and trade. Fish and Aquaculture Technical Papers. 2008;**516**:1-7. Available from: https://www.researchgate.net/publication/281566572

[91] ODSlocal. 2020. Available from: https://odslocal.pt/ [Accessed: November 14, 2020]

[92] Brander K. Impacts of climate change on fisheries. Journal of Marine Systems. 2009;**79**:389-402. DOI: 10.1016/j.jmarsys.2008.12.015. www.acia.uaf.edu

[93] Philippart CJM, Anadón R, Danovaro R, Dippner JW, Drinkwater KF, Hawkins SJ, et al. Impacts of climate change on European marine ecosystems: Observations, expectations and indicators. Journal of Experimental Marine Biology and Ecology. 2011;**400**:52-69. DOI: 10.1016/j.jembe.2011.02.023

[94] Ullah H, Nagelkerken I, Goldenberg SU, Fordham DA. Climate change could drive marine food web collapse through altered trophic flows and cyanobacterial proliferation. PLoS Biology. 2018;**16**(1):e2003446.

DOI: 10.1371/journal.pbio.2003446. [accessed 2020 Oct 22]. Available from: https://pubmed.ncbi.nlm.nih.gov/29315309/

[95] Harley CDG, Hughes AR, Hultgren KM, Miner BG, Sorte CJB, Thornber CS, et al. The impacts of climate change in coastal marine systems. Ecology Letters. 2006;**9**(2):228-241. DOI: 10.1111/j.1461-0248.2005.00871.x. [accessed 2020 Oct 20]. Available from: https://pubmed.ncbi.nlm.nih.gov/16958887/

[96] Brander KM. Global fish production and climate change. Proceedings of the National Academy of Sciences of the United States of America. 2007;**104**(50):19709-19714. DOI: 10.1073/pnas.0702059104. [accessed 2020 Oct 20]. Available from: www.pnas.orgcgidoi 10.1073pnas.0702059104

[97] Philippart CJM, Van Aken HM, Beukema JJ, Bos OG, Cadée GC, Dekker R. Climate-related changes in recruitment of the bivalve *Macoma balthica*. Limnology and Oceanography. 2003;**48**(6):2171-2185

[98] Witbaard R, Duineveld GCA, Amaro T, Bergman MJN. Growth trends in three bivalve species indicate climate forcing on the benthic ecosystem in the southeastern North Sea. Climate Research. 2005;**30**:29-38 www.int-res.com

[99] Chícharo L, Chícharo MA. Effects of environmental conditions on planktonic abundances, benthic recruitment and growth rates of the bivalve mollusc *Ruditapes decussatus* in a Portuguese coastal lagoon. Fisheries Research. 2001;**53**(3):235-250. DOI: 10.1016/S0165-7836(00)00290-3

[100] Beukema J, Dekker R. Decline of recruitment success in cockles and other bivalves in the Wadden Sea: Possible role of climate change, predation on postlarvae and fisheries. Marine Ecology Progress Series. 2005;**287**:149-167. DOI: 10.3354/meps287149. [accessed 2020 Oct 20]. Available from: http://www.int-res.com/abstracts/meps/v287/p149-167/

[101] Beukema JJ, Dekker R, Essink K, Michaelis H. Synchronized reproductive success of the main bivalve species in the Wadden Sea: Causes and consequences. Marine Ecology Progress Series. 2001;**211**(143):143-155

[102] Strasser M, Hertlein A, Reise K. Differential recruitment of bivalve species in the northern Wadden Sea after severe winter of 1995/96 and of subsequent milder winters. Helgoland Marine Research. 2001;**55**(3):182-189. DOI: 10.1007/s101520100080. [accessed 2020 Oct 22]. Available from: http://link.springer.com/10.1007/s101520100080

[103] Baba K, Tada M, Kawajiri T, Kuwahara Y. Effects of temperature and salinity on spawning of the brackish water bivalve *Corbicula japonica*. Marine Ecology Progress Series. 1999;**180**:213-221

[104] Ali AYA, Idris AM, Ebrahim AM, Eltayeb MAH. Brown algae (Phaeophyta) for monitoring heavy metals at the Sudanese Red Sea coast. Applied Water Science. 2017;**7**(7):3817-3824. DOI: 10.1007/S13201-017-0529-1/TABLES/6. [accessed 2021 Nov 26]. Available from: https://link.springer.com/article/10.1007/s13201-017-0529-1

[105] Chandía C, Salamanca M. Long-term monitoring of heavy metals in Chilean coastal sediments in the eastern South Pacific Ocean. Marine Pollution Bulletin. 2012;**64**(10):2254-2260. DOI: 10.1016/J.MARPOLBUL.2012.06.030

[106] Mirzajani A, Hamidian AH, Hassan J. Metal concentrations in the coastal Fauna of the Caspian Sea. Ocean

Science Journal. 2021;**56**(3):256-265. DOI: 10.1007/S12601-021-00023-1/TABLES/3. [accessed 2021 Nov 26]. Available from: https://link.springer.com/article/10.1007/s12601-021-00023-1

[107] Idriss AA, Gao Y-Z, Zhu N-Z, Bizimana A, Meng X-Z. Persistent organic pollutants in polar ecosystems: Current situation and future challenges under climate change. International Journal of Science and Advances. 2021;**2**:295-307. DOI: 10.51542/ijscia.v2i3.12. [accessed 2021 Nov 26]. www.ijscia.com

[108] Miraji H, Ripanda A, Moto E. A review on the occurrences of persistent organic pollutants in corals, sediments, fish and waters of the Western Indian Ocean. Egyptian Journal of Aquatic Research. 2021;**47**(4):373-379. DOI: 10.1016/J.EJAR.2021.08.003

[109] Sanganyado E, Chingono KE, Gwenzi W, Chaukura N, Liu W. Organic pollutants in deep sea: Occurrence, fate, and ecological implications. Water Research. 2021;**205**:117658. DOI: 10.1016/J.WATRES.2021.117658

[110] Bagaev A, Esiukova E, Litvinyuk D, Chubarenko I, Veerasingam S, Venkatachalapathy R, et al. Investigations of plastic contamination of seawater, marine and coastal sediments in the Russian seas: A review. Environmental Science and Pollution Research. 2021;**28**(25):32264-32281. DOI: 10.1007/S11356-021-14183-Z

[111] da Rocha FOC, Martinez ST, Campos VP, da Rocha GO, de Andrade JB. Microplastic pollution in southern Atlantic marine waters: Review of current trends, sources, and perspectives. Science of the Total Environment. 2021;**782**:146541. DOI: 10.1016/J.SCITOTENV.2021.146541

[112] Gassett PR, O'Brien-Clayton K, Bastidas C, Rheuban JE, Hunt CW, Turner E, et al. Community science for coastal acidification monitoring and research. Coastal Management. 2021;**49**(5):510-531. DOI: 10.1080/08920753.2021.1947131. [accessed 2021 Nov 26]. Available from: https://www.tandfonline.com/doi/abs/10.1080/08920753.2021.1947131

[113] Gonski SF, Horwith MJ, Albertson S, Bos J, Brownlee AS, Coleman N, et al. Monitoring ocean acidification within state Borders: Lessons from Washington state (USA). Coastal Management. 2021;**49**(5):487-509. DOI: 10.1080/08920753.2021.1947130/SUPPL_FILE/UCMG_A_1947130_SM5128.DOCX. [accessed 2021 Nov 26]. Available from: https://www.tandfonline.com/doi/abs/10.1080/08920753.2021.1947130

[114] Pashaei R, Loiselle SA, Leone G, Tamasi G, Dzingelevičienė R, Kowalkowski T, et al. Determination of nano and microplastic particles in hypersaline lakes by multiple methods. Environmental Monitoring and Assessment. 2021;**193**(10):1-15. DOI: 10.1007/S10661-021-09470-8/TABLES/1. [accessed 2021 Nov 26]. Available from: https://link.springer.com/article/10.1007/s10661-021-09470-8

[115] Shea R, Schwarzmann D. Proposed Lake Ontario National Marine Sanctuary Study Area Profile. 2021. Available from: https://repository.library.noaa.gov/view/noaa/29627 [Accessed: November 26, 2021]

Chapter 5

Non-Indigenous Marine Fish in Syria: Past, Present and Impact on Ecosystem, and Human Health

Adib Saad and Lana Khrema

Abstract

Biological invasions have posed a major threat to global and regional biodiversity. The Mediterranean, one of the world's main biodiversity hotspots, has long suffered from multiple and recurrent invasions. Due to the geographical location of Syria on the eastern Mediterranean coast, this chapter includes a historical and recent study over the past hundred years of the biodiversity reality of fish fauna in Syrian marine waters (as a representative part of the Levantine Basin). It also includes the evolution non-native fish species number that migrated to this area, both from the Red Sea (through the Suez Canal) and from its original habitats in both Indian and Pacific Ocean, or from the Atlantic Ocean through the Strait of Gibraltar. Then, its spread extended to Syrian territorial waters due to the impact of climate change, which led to environmental changes in the characteristics of the marine waters. An explanation will also be given of the impact of non-indigenous species on native species such as competition for food, ecological niches, and predation, as well as the positive and negative effects on the economy and human health. The consumption of some invasive species, like buffer fish rich in tetrodotoxin, may lead to severe intoxication, sometimes to death.

Keywords: alien species, Lessepsian migration, Mediterranean Sea, tropicalization, marine ecosystems, range extension, Syria

1. Introduction

The habitats of the world's seas and oceans are under siege by invasive alien species, leading to adverse and irreversible consequences for native biodiversity, ecosystem services, the economy, and human health. However, there are some positive aspects resulting from the migration of certain species of high economic value (such as certain types of fish, mollusks, and crustaceans), provided that there is no intense environmental and food competition with local species. Climate change, the increase in the number, tonnage and speed of merchant ships, the widening and deepening of artificial channels, the pollution, and the increase in various human activities contribute to the success of the introduction and establishment of these species. The competition on habitats is a major threat to the structure and functioning of critical marine communities, destroys the exploitable areas, or reduces the productivity of marine environments. In this context, long-term monitoring studies have become necessary to assess the magnitude of these factors and prepare effective management plans to deal with these challenging issues [1]. The Mediterranean, one of the main biodiversity hotspots in the world, has always suffered from multiple and recurrent invasions. The first alien species recorded in the Mediterranean were a pair of fouling serpulid polychaetes, *Hydroids dianthus* Morch, 1863, and *H. diramphus* (Verrill, 1873), collected from the ports of Izmir and Naples in 1865 and 1870, respectively [2, 3]. As the Suez Canal was inaugurated in 1869, we can say that the arrival of these two species in the Mediterranean is due to cargo or transport ships. The next two species were Red Sea mollusks, *Pinctada radiata* (Leach, 1814) and *Cerithium scaridum* Philippi, 1848, collected at Alexandria and Port Said in 1874 and 1883, respectively [4, 5], announcing the Eritrean invasion of the Mediterranean Sea. The influx of alien species has continued since, and evolves with time, human activities, climate change and resulting changes in the physicochemical characteristics of Mediterranean water.

The impacts of alien species have not been undertaken in a serious and deepened way until the beginning of the twenty-first century. The preponderance of species arriving in the Mediterranean *via* the Red Sea (called Eritrean) in the south-east of the Mediterranean, which was called integral Syria (**Figure 1**), or natural Syria [6].

The invasions have been perceived until the beginning of the twenty-first century as singular and widely in the world that bioinvasions constitute one of the most important elements of global changes, with often harmful effects on biodiversity, the economy, and human health.

Many atlases and articles (fish, crustacean and mollusks) summarize existing knowledge of the scale and impact of alien species in the Mediterranean [7, 8]. Research publications, surveys, and conference abstracts indicate that the alien species recorded in the Mediterranean Sea till 2020 more than 188 species of fish are thus

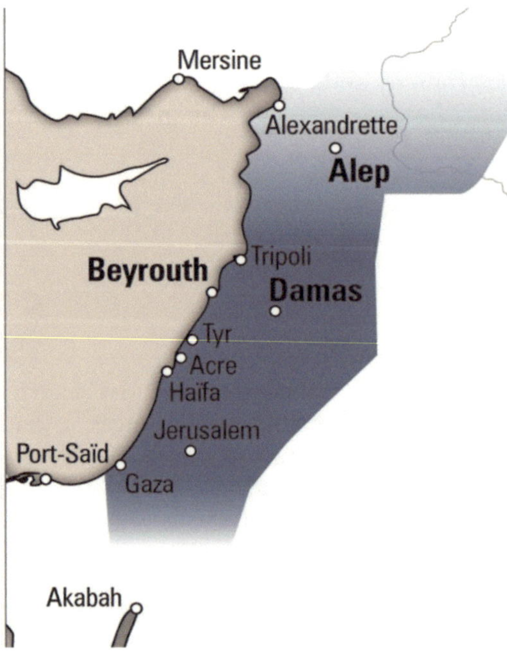

Figure 1.
Map of the coast of greater Syria in 1914, this coast that grovel surveyed marine life during 1929–1931.

broken down: 25 introduced by shipping, mariculture and aquarium releases, or by other human activities; 106 species crossed the Suez Canal; 57 Atlantic species whose arrival in the Mediterranean Sea is attributed to unassisted migration through the Strait of Gibraltar [9]. Through our current reference study, we estimate the number of fish species that entered the Mediterranean until the end of 2022 at about 214 species, of which about 120 species entered through the Suez Canal, 65 species of Atlantic species, and 29 species introduced by shipping, aquaculture, or other human activities. We will focus in this chapter on the group of exotic fish invading the territorial waters of (current) Syria, which constitute the extreme east of the Mediterranean coast and a hot spot to receive species introduced into the Mediterranean Sea *via* the Suez Canal. This is attributed to the great Atlantic current from west toward the east (Coriolis Force), bringing with it the larvae, eggs, and adults preferably toward the east and north in front of the east coast, before gradually scattering in the eastern basin, and then throughout the Mediterranean (**Figure 2**).

Figure 2 clarifies that the eastern coast of the Mediterranean, with the Syrian coast in the middle, is primarily subject to the arrival of migratory species from the Red Sea (Lessepsian), due to the main marine current heading from east to west and then to the north in a counterclockwise direction, in addition to the environmental characteristics (temperature and salinity), closer to the characteristics of the waters of the Red Sea than the rest of the Mediterranean.

Syrian sea water occupies about 4000 km^2 of the Levantine Basin in the eastern part of the Mediterranean, where the Syrian coast extends 183 km from the Turkish border in the north to the Lebanese border in the south (**Figure 3**). The Syrian marine waters as part of the Levantine Basin and the easternmost part of the Mediterranean Sea are characterized by high salinities and high temperature. In the Mediterranean, salinity exhibits an east ward increase, from approximately 37.5‰ in the west to 39.5‰ in the east; temperature increases from west to east, ranging from 15 to 26°C [11]. Gruvel [6] was the first specialist to report on the marine ichthyofauna of the Eastern

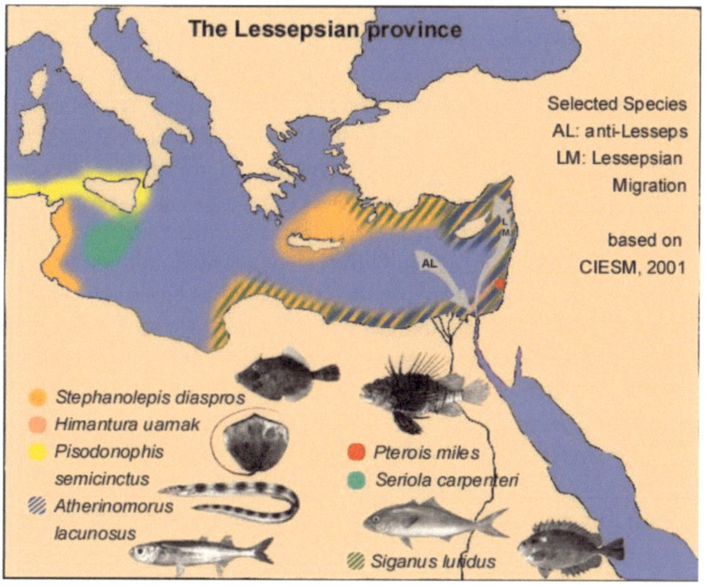

Figure 2.
Species migration from the Red Sea to eastern coast of the Mediterranean.

Figure 3.
The exclusive economic zone (EEZ) and shelf area (to 200 m depth) of Syria [10].

Mediterranean Sea. Forty years later, a local study was carried out on the marine ichthyofauna from the Syrian coast [12]. During the last three decades, several studies have been carried out sporadically, and all recorded species have been documented with voucher specimens, photographs, and scientific publications [13–17].

2. Definitions and working methods

An alien species is defined as an organism present outside its known or consensual range [18]. Alien species were grouped into four main categories, namely established, casual, questionable, and cryptogenic, the definition of these terms can be found in Ref. [19]. Clearly and concisely, the established species are the exotic species with self-maintaining populations; casual species are the species which were reported only once in the region; questionable species are the species which were reported without sufficient information and their taxonomic status is uncertain; cryptogenic species are the species with no definite evidence of their native or introduced status according to Carlton [20]. Other important definitions to know are as follows:

Invasive species are the established exotic that have overcome biotic and abiotic barriers in the region and are able to expand their distributional ranges through the production of fertile offspring, with noticeable impact on the prevailing ecosystem.

The Lessepsian migration refers to the migration through the Suez Canal (planned by Ferdinand de Lesseps), generally, from the Red Sea to the Mediterranean, and rarely in an opposite way, favored by two distinct man-made events: the creation of a waterway between the Indo-Pacific and the Mediterranean basins and the control of the Nile fresh-water outflow.

Range expansion refers to species that have recently arrived in a given area by natural dispersal from a neighboring area in which they are native, without the intentional or unintentional intervention of human.

The increase in shipping and the opening of the Suez Canal predates modern studies of Mediterranean marine taxa, saving mollusks and fish, by at least half a century. Since the beginning of the twentieth century, extensive biological studies have been carried out in the Mediterranean, allowing a reasonable measure of confidence in the separation of alien from native biota in the best-known taxa. Since the probability of encountering a stray animal at sea is less and less important, most recorded alien species are considered "established" species that have self-sustaining populations of some duration. It is recognized that some alien species may not maintain their populations over time, and therefore, a single record dating back decades may be considered an ephemeral entry. The distinction between "established" and "ephemeral" extraterritorials can vary in space and time and is sometimes difficult to discern and largely circumscribed by our unfamiliarity with the marine environment. This study is geographically restricted to the Levantine Basin—in general—and specifically the Syrian marine waters. The list of species was taken from several sources. Primary sources include research articles, fish fauna inventories, fisheries management studies and reports, author-supervised, and unpublished M.Sc. and Ph.D. theses conducted over the past 30 years. The dataset includes the native range of the extraterritorial, and its means of introduction, whether through the Suez Canal, ships (drilling hull transport, fouling, crevicultural or clinging species, and ballast), marine cultivation (intentional or not), and other commercial introductions (ornamental, bait, edible species).

The results presented in this chapter depend mainly on the results of field research conducted by the author and his team during the past three decades, represented by the results of master's and doctoral researches and the author's research whose results have been published in scientific journals. In addition, the data for this chapter included a review of the results of work published by other researchers in the area. The work included following up the landing of fish in fishing ports, studying the quantitative and qualitative composition of the catch, as well as following up with local fishermen to inform the author of any unfamiliar fish species so that the author or a member of his team could bring the fish sample to the laboratory for study and classification, and then the results in a scientific journal are published. Each reference will be mentioned next to each new result in succession in the results and discussion section.

3. Results and discussion

3.1 List of species and composition of fish fauna

A total of 106 species of exotic fish have been identified in Syrian marine waters until end of August 2023. While the number in 1931 was only seven species [6], in 1976 it rose to 11 species [12], while in 2005 it reached 54 species [15], and since that date the acceleration in recording the presence of exotic species has now reached 106 species out of 343 species (**Tables 1** and **2**). The total number of fish species in Syrian marine waters, that is, about 30% of the total fish fauna in this region, and this large increase in the number of alien species are attributed to the following:

1. climate changes and the subsequent changes in the environmental characteristics of water and the food chain,

Category	Species	Families	Orders	Alien
Chondrichthyes	50	25	10	5
Actinopterygii	292	98	26	101
Class:Hyperoartia	1	1	1	—
Total	343	124	37	106

Table 1.
Number of recorded fish species and lamprey from Syrian Marin water, and number of alien species.

2. the increase in human activities represented by fishing and transportation,

3. the second Suez Canal was built, which increased the migration of more tropical species from the Red Sea to the Mediterranean (**Figure 3**), and

4. increased monitoring and scientific research in the field of biodiversity and ecosystems.

3.2 Coexistence with local species

Alien species composition differs between ecoregions in the Mediterranean; studies have shown that these differences are greater for species introduced by Lessepsian and for aquaculture species. The spatial pattern of biodiversity of native and exotic species is different: The overall richness of native species decreases from northwestern to southeastern regions, while the opposite trend is observed for exotic species [19].

The impact of the exotic migration was enormous on the ecosystem of the Levant Basin, within which the Syrian marine waters lie. The fact that many species have proven successful in the new environment indicates that the nutritional requirements of many migratory species reflect the non-selective nature of their feeding habits. This adaptation is of great importance to any immigrant in its new environment. It is self-evident that successful colonization can only be established when the overlap between the environmental conditions in the source and target regions is within the colonizer's tolerance (**Figure 4**). Since these environments are not identical, the colonizing population will respond to new selective pressures by moving away from the parent population, thus becoming more adapted to their new habitat. The successful establishment of the large number of exotic species raises the question of complex interactions with native species. However, this high number is determined mainly by the ability of the migrants to occupy a niche in the new ecosystem, maintain itself by successful breeding when environmental and hydrological conditions are suitable for reproduction and growth of larvae to maturity, and then replenish and develop their stocks in the new environment.

3.3 Impact of alien species on marine ecosystems

The impact of alien species, known as biological contamination, is something that must be considered in environmental impact assessments [81]. Human-mediated introduction of marine species to a region outside their natural range of distribution is widely considered to be the major threat to indigenous species diversity and community structures. This can cause habitat modifications, changes in ecosystem functioning, introduction of new diseases and parasites, and genetic modifications, such as

Family	Species	Origin	Status	First record	Ref.
Apogonidae	*Apogon atradorsatus*	South-east pacific	sin	Alshawy et al., 2019a	[21]
	Apogonichthyoides pharaonic (mis id Apogon taeniatus)	Red sea Indo-Pacific	est	Sbaihi and Saad, 1992	[22]
	Cheilodipterus novemstriatus	Indian	est	Ali et al., 2018	[23]
	Epigonus constanciae	Western Mediterranean	N	Sbaihi, 1994 (as *Epigonus telescopus* (synonyme))	[13]
	Jaydia smithi	Red sea	sin	Alshawy et al., 2017	[24]
	Jaydia queketti	Western Indian Ocean	sin	Al shawy et al., 2019b	[25]
	Ostorhinchus fasciatus	Indo-West Pacific	sin	Al Shawy et al., 2019c	[26]
Argentidae	*Aregentina sphyraena*	Western Mediterranean		Saad and Sbaihi, 1995	[27]
	Glassanodon leioglossus	Western Mediterranean		Saad and Sbaihi, 1995	[27]
Atherinidae	*Atherinomorus forskalii (=Atherinomorus lacumosus)*	Red sea Indo-Pacific	est	Saad et al., 2002	[28]
	Atherinomorus lacunosus	Indo-Pacific	sin	Othman et al., 2022	[29]
	Atherinomorus pinguis (=Pranesus pinguis)	Red sea		Saad et al., 2002	[28]
Belonidae	*Ablennes hians*	Worldwide distribution	sin	Al Shawy et al., 2019d	[30]
	Tylosurus crocodilus (=Tylosrus corma)	Red sea Indian	est	Saad et al., 2002	[28]
Blenniidae	*Petroscirtes ancylodon*	Red sea Indian	est	Saad, 2002	[31]
Bramidae	*Brama brama*	Western Mediterranean	N	Saad and Sbaihi, 1995	[27]
Bregmacerotidae	*Bregmaceros nectabanus*	Atlantic and Indian Ocean and tropical western Pacific	sin	Othman & Galiya, 2019	[32]
Callionymidae	*Callionymus filamentosus*	Red sea Indo-Pacific	est	Saad and Sbaihi, 1992	[33]
Caproidae	*Capros aper*	Western Mediterranean	N	Saad and Sbaihi, 1995	[27]
Carangidae	*Aleps djedaba*	Red sea Indo-Pacific	est	Saad, 2002	[31]
Carangidae	*Naucrates ductor*	Eastern Atlantic and Indian Ocean	est	Ali-Basha et al., 2021	[34]
Chaetodontidae	*Chaetodon larvatus*	Indian	sin	Ali et al., 2017a	[35]
	Heniochus intermedius	Indo-West Pacific Ocean	sin	Saad et al., 2022a	[36]
Champsodontidae	*Champsodon nudivittis*	Indo-Pacific	est	Ali et al., 2017b	[37]

Family	Species	Origin	Status	First record	Ref.
Clupeidae	*Dussumieria elopsoides*	Red sea Indo-Pacific	est	Saad, 2002	[31]
	Etrumeus sadina (=Eutrimeus teres)	Red sea Indian	est	Saad, 2002	[31]
	Herklotsichthys punctatus	Red sea Indo-Pacific	est	Saad, 2005	[15]
Cynoglossidae	*Cynoglossus sinusarabici*	Red sea Indo-Pacific	est	Saad and Sbaihi, 1992	[33]
Dussumieriidae	*Dussumieria elopsoides*	Red sea Indo-Pacific	est	Saad, 2002	[31]
	Etrumeus sadina (=Eutrimeus teres)	Red sea Indo-Pacific	est	Saad, 2002	[31]
Epigonidae	*Epigonus denticulatus*	Western Mediterranean		Ibrahim et al., 2023	[38]
Exocoetidae	*Parexocoetus mento*	Red sea Indo-Pacific	est	Saad et al., 2002	[28]
Fistulariidae	*Fistularia commersonii*	Red sea Indo-Pacific	est	Saad, 2002	[31]
	Fistularia petimba	Atlantic and Indo-Pacific	sin	Hussein et al., 2019	[39]
Gadidae	*Micromesistius poutassou*	Western Mediterranean	N	Saad and Sbaihi, 1995	[27]
	Phycis phycis	Western Mediterranean	N	Saad and Sbaihi, 1995	[27]
	Gadiculus argenteus	Western Mediterranean	N	Saad and Sbaihi, 1995	[27]
Gobiidae	*Oxyurichthys petersi*	Red sea	est	Saad and Sbaihi, 1992	[33]
	Silhouetta aegyptia	Red sea Indian	est	Sbaihi and Saad, 1995	[27]
Gobiesocidae	*Lepadogaster lepadogaster*	Western Mediterranean	N	Saad and Sbaihi, 1995	[27]
	Lepadogaster candolli	Western Mediterranean		Saad and Sbaihi, 1995	[27]
Haemulidae	*Pomadasys stridens*	Indian	est	Saad, 2005	[15]
Hemiramphidae	*Hemiramphis far*	Red sea Indo-Pacific	est	Gruvel, 1931	[6]
	Hyporamphus affinis	Red sea Indo-Pacific	est	Saad, 1996	[39]
Heterenchelydae	*Panturichthys flowleri*	Western Mediterranean		Sbaihi, 1994	[13]
Holocentridae	*Sargocenrum rubrum*	Red sea Indo-Pacific	est	Anon, 1976	[12]
Labridae	*Pteragogus trispilus*	Indian	est	Soliman et al., 2014	[40]
	Symphodus bailloni	Eastern Atlantic	sin	Khrema et al., 2022	[41]

Family	Species	Origin	Status	First record	Ref.
Leiognathidae	*Equulites klunzingeri* (= *Leiognathus kluningeri*)	Red sea Indo Pacific	est	Sbaihi and Saad, 1995	[27]
	Equulites popei	Indo-West Pacific	sin	Ibrahim et al., 2020	[42]
	Leiognathus berbis	Indo-Pacific	est	Alshawy et al., 2016	[43]
Lophotidae	*Lophotus lacepede*	Western Mediterranean		Ali et al., 2021	[44]
Lutjanidae	*Lutjanus fulviflamma*	Indo-Pacific:	sin	Saad et al., 2022	[45]
Monacanthidae	*Stephanolepis diaspros*	Red sea Indian	est	Gruvel, 1929	[6]
Mugilidae	*Liza carenata*	Red sea Indian	est	Saad, 1995	[46]
Mullidae	*Parupeneus forsskali*	Indian	est	Ali et al., 2016b	[47]
	Parupeneus rubescens	Indo-West Pacific		Sabour and Masri, 2022	[48]
	Upeneus moluccensis	Red sea Indo-Pacific	est	Gruvel, 1931	[6]
	Upeneus pori	Red sea	est	Sbaihi and Saad, 1995	[27]
Nemipteridae	*Nemipterus randalli*	Indian	est	Ali et al., 2013	[49]
Ophichthidae	*Ophisurus serpens*	Atlantic and Indo-Pacific	sin	Al Shawy et al., 2019d	[50]
Ophidiidae	*Dalophis imberbis*	Eastern Atlantic	rex	Capape et al., 2021	[51]
	Ophidion rochei	Mediterranean and Black Sea	rex	Othman et al., 2020	[52]
	Ophidion rochei	Mediterranean and Black Sea	rex	Othman et al., 2020	[52]
Ostraciidae	*Tetrosomus gibbosus*	Red sea Indo-Pacific	est	Saad, 2002	[31]
Pempheridae	*Pempheris rhomboidea* *Pempheris vanicolensis*	Red sea Indo-Pacific	est	Sbaihi and Saad, 1992 as P. vanicolensis	[22]
Platycephalidae	*Platycephalus indicus*	Red sea Indo-Pacific	est	Saad et al., 2002	[28]
Plotosidae	*Plotosus lineatus*	Indo-Pacific	est	Ali et al., 2015	[53]
Pomacanthidae	*Pomacanthus imperator*	Indo-Pacific	sin	Saad et al., 2018	[54]
	Pomacanthus maculosus	Red sea Western Indian Ocean		Capapé et al., 2023	[55]
Pomacentridae	*Abudefduf vaigiensis*	Indo-Pacific	sin	Saad et al., 2020a	[56]
Priacanthidae	*Priacanthus hamrur*	Indo-Pacific		Capape et al., 2022	[57]
	Priacanthus sagittarius	Indo-West Pacific	sin	Al Shawy et al., 2019e	[58]
Scaridae	*Scarus ghobban*	Indo-Pacific	est	Saad et al., 2018	[59]
Scombridae	*Scomberomorus commerson*	Red sea Indo-Pacific	est	Anon, 1976	[12]
	Scomber indicus	Indo-Pacific	sin	Othman et al., 2023	[60]

Family	Species	Origin	Status	First record	Ref.
Scorpaenidae	*Pterois miles*	Indian	est	Ali et al., 2016	[61]
	Pterois volitans	Indo-Pacific		Fandi et al., 2022	[62]
Serranidae	*Epinephelus areolatus*	Indo-Pacific	2 records	Al Mabruk et al., 2021	[63]
	Epinephelus caninus	Eastern Atlantic	est	Saad et al., 2020b	[64]
	Epinephelus fasciatus	Red sea	sin	Foulquie & Dupuy de la Grandrive, 2003	[65]
Siganidae	*Siganus javus*	Indo-Pacific	sin	Ibrahim et al., 2010	[66]
	Siganus luridus	Red sea Indian	est	Gruvel, 1931	[6]
	Siganus revulatus	Red sea	est	Gruvel, 1931	[6]
Soleidae	*Pegusa lascaris (=Solea lascaris)*	Western Mediterranean	N	Sbaihi, 1994	[13]
Sparidae	*Acanthopagrus bifasciatus*	Western Indian Ocean	sin	Saad et al., 2022b	[67]
	Crenidens crenidens	Red sea Indo-Pacific	est	Saad et al., 2002	[28]
	Pagellus bellottii	western Mediterranean	N	Sbaihi and Saad, 1992	[68]
	Pagellus bogaraveo	Eastern Atlantic and west Mediterranean	sin	Saad et al., 2020	[69]
	Pagrus major	Northwest Pacific	sin	Saad et al., 2022	[70]
	Rhabdosargus haffara	Red Sea	est	Saad, 2005	[15]
	Rhabdosargus sarba	Indo-West Pacific	sin	Hamwi and Ali-Basha, 2021	[71]
Sphyraenidae	*Sphyraena chrysotaenia*	Red sea Indo-Pacific	est	Saad, 2002	[31]
	Sphyraena flavicauda	Red sea Indo-Pacific	est	Saad et al., 2002	[28]
Synanceiidae	*Synanceia verrucosa*	Indo-Pacific	sin	Ibrahim et al., 2019	[72]
Syndontidae	*Saurida lessepsianus Saurida undosquamis*	Red sea Indo-Pacific	est	Anon, 1976 as S. undosquamis	[12]
Tetraodontidae	*Lagocephalus guentheri (=Lagocephalus spadiceus)*	Red sea	est	Anon, 1976 as L. spadiceus	[12]
	Lagocephalus lagocephalus	Circumglobal		Alshawy et al., 2019c	[73]
	Lagocephalus sceleratus	Indo-Pacific	est	Khalaf et al., 2014	[74]
	Lagocephalus suezensis	Red sea	est	Saad et al., 2002	[28]
	Torquigener flavimaculosus	Indian	est	Sabour et al., 2014	[75]
Theraponidae	*Pelates quadrilineatus*	Red sea Indo-Pacific	est	Saad, 2005	[15]
	Terapon puta	Red sea Indo-Pacific	est	Saad, 2005	[15]

Family	Species	Origin	Status	First record	Ref.
Elasmobranchii:					
Dasyatidae	*Himantura uarnak*	Indo-Pacific	est	Ali et al., 2010	[76]
	Himantura leoparda	Indo-West Pacific	sin	Saad et al., 2021a	[77]
Rajidae	*Leucoraja circularis*	Eastern Atlantic	sin	Alkusairy and Saad, 2018	[78]
	Leucoraja fullonica	Eastern Atlantic	sin	Saad and Alkusairy, 2019	[79]
Torpedinidae	*Torpedo sinuspersici*	Indo-Pacific	cas	Saad et al., 2004	[14]
Jawless fishes (Class: Hyperoartia)					
Petromyzonidae	*Petromyzon marinus*	Northwestern Atlantic	sin	Saad et al., 2021b	[80]

Table 2.
Non endogenous fish in Syrian marine water, category, native range, and first records. Species in alphabetic order within each class and family. (Est = established, sin = single record, case = casual, rex = rang extending).

Figure 4.
Red Sea goatfish Parupeneus forsskali *that was recorded for the first time in the Syrian coast in 2015 [47]; it was able, within only 5 years, to become abundant in the catch with gill nets instead of the rest of the species of the mulledi family, and this is one of the examples of the competition of invasive species with indigenous species.*

hybridization with the native taxa [82], rabbitfishes (siganidae) have by now overgrazed a significant part of the eastern Mediterranean coast (including Syrian coast), degrading rich habitats of brown algal forests into bare rock, and such impacts, including completion with native herbivores, are likely to be exacerbated by climate change. In several instances, exotic species will outcompete native species, such as *Fistularia commersonii* and *Pterois miles*, and prey on various fish and invertebrates—some of commercial importance with impacts on both local population and fisheries. Our field study showed that *F. commersonii* is carnivorous and feed on many small fish species, such as *Sardinella* spp., *Sprarus spratus*, *Alosa fallax*, *Boops bops*. About 12–30 individual preys were found in the stomach of every specimen (**Figure 5**). The high fecundity rate, and its rapidly expanding population feeding on fish and some invertebrates, makes a decline of the biomass and causing damage and loss to fisheries yield [83, 84]. On the other hand, the results of a survey study during the years 2013–2014

Figure 5.
The content of the digestive tube of one individual of the fish F. commersonii, *showing 22 individuals of prey belonging to several endemic species of small pelagic fish (and a greater number of these prey were found in other samples in several cases), indicating the danger of this exotic species to native fish fauna, biodiversity, and local fish stocks.*

showed that exotic fish (most of them are Lessepsian) constitute more than 27% of the weight catch by artisanal fishing (means gill net and trammel net) and more than 50 of trawl net in Syria [83, 84]. Comprehensive evaluation of the overall consequences of alien introductions must integrate the impact of climate change, which clearly reduces the suitability of the eastern Mediterranean waters (including Syrian waters) to several native fish species. This is already observed in this area, where no indigenous species of tropical origin are replacing the ecological functions of temperate native biota that are compromised by overfishing.

3.4 Impact of non-indigenous species on the public health

Fish invasions in many cases impact human health, and one of the best documented examples is provided the buffer fish *Lagocephalus scelertaus*, which has caused severe damages since its relative recent apparent in Syrian marine waters. The consumption of this invasive species, rich in tetrodotoxin, may lead to severe intoxication, sometimes to death. By examining hospital records in both Lattakia and Tartous during the past 10 years, it is clear that this fish has caused many cases of poisoning and several cases of death for consumers in Syria [85–87].

The marketing of this species is currently banned in Syria and in other Mediterranean countries. To inform local populations about the risks associated with this invasion, Syrian authorities and many non-governmental associations (NGO) have launched awareness campaigns (**Figures 6–8**), which have succeeded in reducing the risk of inadvertent consumption by obtaining information on the occurrence and abundance of this "unwanted guest." Other dangerous species include *Pterois miles*, *Plotosus lineautus*, *Synannceia verrucosa*, and two rabbitfishes (*Siganus luridus* and *Siganus rivulatus*) all possessing venomous spines and capable of causing severe injuries [88]. The impact of invasive fishes is not all negative, however. Often food provision is a clear benefit of invasions. That is the case of rabbitfishes, goatfishes (*Upeneus moluccensis, Upeneus pori*, *Parupeneus forsskali*), and other edible species such

Figure 6.
Example of campaign signaling the toxicity of the silver-cheeked toadfish (Lagocephalus sceleratus) *in Syria (by Syrian society for aquatic environment protection (NGO) and general commission of fisheries and aquatic resources). The text illustrates how to distinguish from native pufferfishes. The poster illustrates the risk associated with the consumption of this toxic species.*

Figure 7.
Example of campaign signaling the poisoning through prickling of strpeed eel catfish (Plotosus lineatus, *(by Syrian society for aquatic environment protection (NGO)).*

as *Nemipterus randelli, Saurida lessepsianus*, and *Scomberomorus commerson*, which have over years developed abundant populations in the Leventine basin, to the point of becoming a target for local fisheries [10, 83].

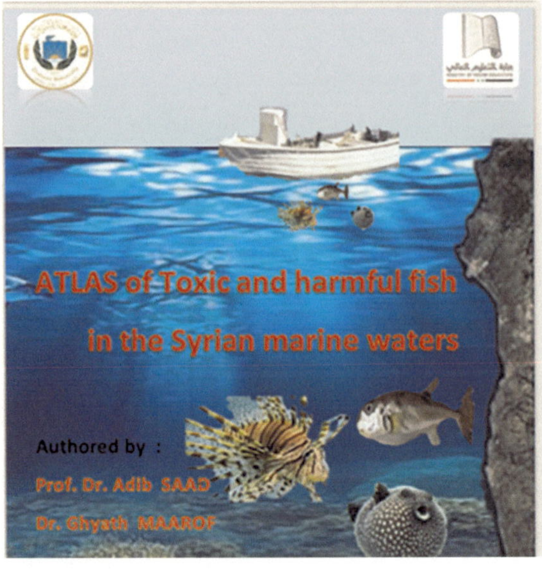

Figure 8.
The cover of the Atlas of toxic and harmful fish that threaten human health and community safety, which was published by Tishreen university and the Syrian Society for the Protection of the aquatic environment, addressed to the local population in Arabic [88].

A scientific explanation was provided for each hazard or damage, including toxins contained in certain species. The research also included illustrations of each fish species, explaining the harmful parts of the body, explaining the level of damage and danger caused by each fish species, as well as an explanation of how to prevent damage and the means of treating each type of damage: /9/ poisonous species in their flesh (i.e., poisoning of humans by eating their meat), /3/poisonous species of blood), and /13/ harmful and poisonous fish species through prickling or stinging poisonous thorns, and/ 4/ types of harmful cartilaginous fish through poisonous stinging with thorns on their body, especially the tail. In addition, / 4 / species of electric fishes that live in Syrian waters were documented (**Figure 8**) [88].

4. Conclusion

The present work elucidates that the marine waters of Syria, as a part of the Levantine basin, are under heavy invasion by alien species, mostly coming from the Red Sea and Indo-Pacific areas. Scientific and public interest in the species introductions and their impacts have resulted in the productions of new papers and synthesis on them and made great contributions to our understanding of the dimensions of this phenomenon. Therefore, this up-to-date inventory of the alien species on the coasts of Syria, apart from its scientific merits, can fulfill the needs of the regulatory requirements and environmental management options for decision makers. Unfortunately, strict precautions and regulations to limit the spread of invasive species have not been implemented in Syria.

It is necessary to consider the classification of non-indigenous and modern species, because these species, which differ in terms of entry pathways, also differ from each other in the ecological, social, and economic effects that they cause in the new environment (predatory fish and competition in the ecological habitat), as well as their impact on the community health of the population, locals, fishermen, and consumers (poisonous and stinging fish). These differences have been demonstrated by ecological and biological studies and practical solutions in the literature. In addition, these differences will directly affect the precautions taken in the sustainable management of the ecosystem and will therefore affect the success of these precautions.

Acknowledgements

The authors thank IntechOpen for helping us find funding to cover publishing expenses. We also thank Dr. Ilene Mahfoud from the Syrian Society for Aquatic Environment Protection (SSAEP) for her efforts in linguistic revision and in preparing figures and maps.

Conflict of interest

The authors declare no conflict of interest.

Author details

Adib Saad[1*] and Lana Khrema[2]

1 Directorate of Scientific Research, Manara University, Lattakia, Syria

2 Faculty of Agriculture, Tishreen University, Lattakia, Syria

*Address all correspondence to: adibsaad52@gmail.com; manara.edu.sy

References

[1] Yapici S. A risk screening of potential invasiveness of alien and Neonative marine fishes in the mediterranean sea: Implications for sustainable management. Sustainability. 2021; **13**(24):13765. DOI: 10.3390/su132413765

[2] Carus JV, 1889–1893. Prodromus Faunae mediterraneae sive descriptio animalium maris Mediterranei incolarum, 2: Brachiostomata. Mollusca. Tunicata. Vertebrata. E. Schweitzerbart, Stuyygart [pp. 1–272 issued 1889]

[3] Zibrowius H. Serpulidae (Annelida Polychaeta) des côtes ouest de l'Afrique et des Archipels Voisins. Annales Musee Royal de l'Afrique Centrale. 1973;**207**:1-93

[4] Monterosato di TA. Enumerazione e sinonimia delle Conchiglie mediterranee. Giornale Di Scienze Naturali Ed Economiche. 1878;**13**:61-115. [December 31st, 1878]. Available from: https://www.molluscabase.org/aphia.php?p=sourcedetails&id=40030 source: MolluscaBase

[5] Keller C. Die Fauna im Suez-Kanal und die Diffusion der Mediterranean und Erythräischen Tierwelt. Neue Denkschriften der allgemeinen Schweizerischen Gesellschaft für die gesamten naturwissenschaften, ser. 1883;**3**(28):1-39

[6] Gruvel A. Les etats de Syrie, richesses marines et fluviales. Paris: Exploitation actuelle et venire. Soc. Edit. Geogr. Marit.et colon; 1931. p. 453

[7] Golani D, Orsi-Relini L, Massuti E, Quignard J-P. In: Briand F, editor. CIESM Atlas of Exotic Species in the Mediterranean. Vol. 1. Fishes. Monaco: CIESM Publications; 2002. p. 254

[8] Golani D, Azzurro E, Dulčić J, Massutí E, Orsi-Relini L. In: Briand F, editor. Atlas of Exotic Species in the Mediterranean Sea. Paris, Monaco: CIESM Publishers; 2021. p. 365

[9] Azzurro E, Smeraldo S, Minelli A, D'Amen M. ORMEF: A Mediterranean database of exotic fsh records. Nature Research, Scientific Data. 2022;**9**(1):363. DOI: 10.1038/s41597-022-01487-z

[10] Ulman A, Saad A, Zylich K, Pauly D, Zeller D. Reconstruction of Syria's fisheries catches from 1950–2010: Signs of overexploitation. Acta Ichthyologica Et Piscatoria. 2015;**45**(3):259-272

[11] Skliris N. Past, present and future patterns of the thermohaline circulation and characteristic water masses of the Mediterranean Sea. In: Goffredo S, Dubinsky Z, editors. The Mediterranean SeaPublisher. Netherlands: Springer; 2014. DOI: 10.1007/978-94-007-6704-1-3

[12] Anon. Commercial fish species collected by the Korean mission for scientific cooperation in Syria local report. Ministry of Agriculture. 1976:76

[13] Sbaihi M. Biosystematic Study of Teleostien Fishes in Marine Water of Syria (Eastern Mediterranean). MSc Thesis. Latakia: Faculty of Sciences, Tishreen University; 1994. p. 220

[14] Saad A, Seret B, Ali M. Liste commentée des Chon- drichthyens de Syrie (Méditerranée orientale). Rapport Commission international Mer Méditerranée. 2004;**37**:430

[15] Saad A. Checklist of bony fish collected from the coast of Syria. Turkish Journal of Fisheries and Aquatic Sciences. 2005;**5**:99-106

[16] Ali M. An updated checklist of marine fishes from Syria with an

emphasis on alien species. Mediterranean Marine Science. 2018; **19**(2):388-393P. DOI: 10.12681/mms.15850

[17] Saad A, Al Kusairy H. Atlas of Sharks and Rays in the Syrian Marine Waters. 2022. Available from: https://www.researchgate.net/publication/362858123

[18] Occhipinti-Ambrogi A, Galil BS. A uniform terminology on bioinvasions: A chimera or an operative tool? Marine Pollution Bulletin. 2004;**49**: 688-694

[19] Zenetos A, Gofas S, Verlaque M, Çinar ME, Raso G. Alien species in the Mediterranean Sea by 2010. A contribution to the application of European Union's marine strategy framework directive (MSFD). Part I. Spatial distribution. Mediterranean Marine Science. 2010;**11**(2):381-493

[20] Carlton JT. Biological invasions and cryptogenic species. Ecology. 1996;**77**: 1653-1655

[21] Alshawy F, Ibrahim A, Hussein C, Lahlah M. First record of the blacktip cardinalfish Apogon atradorsatus Heller & Snodgrass, 1903 from Syrian marine waters (eastern Mediterranean). International Journal of Advanced Research in Science Engineering and Technology. 2019;**6**(3):8299-8302

[22] Sbaihi M, Saad A. Données préliminaires sur la présence d'une espèce de la famille des "Pempheridae" sur la cote de Syrie (Méditerranée orientale). Rapport du XXXIIIe Congrcs CIESM. 1992;**34**:311

[23] Ali M, Khaddour M, Capape C. Confirmed occurrence of Indian Ocean twospot cardinalfish *Cheilodipterus novemstriatus* (Osteichthyes: Apogonidae) in the Levantine Sea: First record from the Syrian coast. Acta Adriatica. 2018;**59**(1):153-158

[24] Alshawy F, Lahlah M, Hussein C. First record of the Lessepsian migrant Smith's cardinalfish Jaydia smithi Kotthaus, 1970 (Pisces: Apogonidae) from Syrian marine waters. Basrah Journal of Agricultural Sciences. 2017; **30**(2):45-49

[25] Alshawy F, Ibrahim A, Hussein C, Lahlah M. First record of the spotfin cardinal fish Jaydia queketti (Gilchrist, 1903) (Teleostei: Apogonidae) from the Syrian marine waters (eastern Mediterranean). Iranian Journal of Ichthyology. 2019;**6**(2):138-142

[26] Alshawy F, Ibrahim A, Hussein C, Lahlah M. First record of the Broadbanded cardinalfish Ostorhinchus fasciatus (white, 1790) from the Syrian marine waters (eastern Mediterranean). SSRG International Journal of Agriculture & Environmental Science (SSRG-IJAES). 2019;**6**(3):14-16

[27] Saad A, Sbaihi M. Mise en évidence de la présence de 11 espèces de téléostéens originaires de l'océan Atlantique et de la méditerranée occidentale dans les eaux de Syrie (méditerranée orientale). Rapp. Comm. int. mer. Medit. 1995;**34**:256

[28] Saad A, Hureau JC, Hammoud V, Ali M. Fish biodiversity and the Impact of environmental factors and human activities in Syrian marine waters. In: Proceeding of 9th conference of Arab Union Biologists. Aleppo; Arab Union Biologists. Cairo, Egypt. 2002. pp. 1-6

[29] Othman R, Galiya M, Almajid Z. First description of Hardyhead silverside *Atherinomorus lacunosus* (Forster in Bloch and Schneider, 1801) from Syrian

marine waters-eastern Mediterranean. Species. 2022;**23**(71):173-177

[30] Alshawy F, Ibrahim A, Hussein C, Lahlah M. First record of the flat needlefish Ablennes hians (Valenciennes, 1846) from Syrian marine waters (eastern Mediterranean). Marine Biodiversity Records. 2019;**12** (15):1-4. DOI: 10.1186/s41200-019-0174-5

[31] Saad A. Characterization of Lessepsian Migrant Fish at Syrian Sea Waters. In: "Mediterranean Vermitid Terrace and Migratory/Invasive Organisms" INOC and SNRSL, Beirut/ Lebanon, act of workshop. 2002.

[32] Othman RM, Galyia M. The appearance of Smallscale Codlet, *Bregmaceros nectabanus* Whitley, 1941 (Bregmacerotidae) in Syrian marine waters. In: Stern N et al., editors. New Mediterranean Biodiversity Records 2019. Vol. 20. Mediterranean Marine Science. Greece: Hellenic Centre for Marine Research (HCMR); 2019. pp. 409-426. DOI: 10.12681/mms.20602

[33] Saad A, Sbaihi M. Etude preliminaires sur un GOBIDAE 'signalé pour la première fois sur la cote de Syrie (Mediterranee orientale). Rapport du XXXIIIe Congres CIESM. 1992;**34**:310

[34] Ali-Basha N, Saad A, Hamwi N, Tufahha A. First record pilotfish Naucrates doctor (Linnaeus 1758), Carangidae, in the Syrian marine waters (Levantine Basin). Marine Biodiversity Records. 2021;**14**(7):1-4. DOI: 10.1186/ s41200-021-00202-y

[35] Ali M, Saad A, Soliman A, Rafrafi-Nouira S, Capape C. Confirmed occurrence in the Mediterranean Sea of Red Sea orange face butterflyfish *Chaetodon larvatus* (Osteichthyes: Chaetodontida) and first record from the Syriancoast. Cahiers de Biologie Marine. 2017;**58**:367-369

[36] Saad A, Masri M, Soliman A, Khrema L. First occurrence of Heniochus intermedius Steindachner, 1893 in the Syrian marine waters (Levantine Basin). Asian Journal of Fisheries and Aquatic Research. 2022;**20**(1):14-18P. DOI: 10.9734/AJFAR/2022/v20i130486

[37] Ali M, Saad A, Jabour R, Rafrafi-Nouira S, Capapé C. First record of nakedband gaper Champsodon nudivittis (Osteichthyes: Champsodontidae) off the Syrian coast (eastern Mediterranean). Journal of Ichthyology. 2017;**57**(1): 161-163

[38] Ibrahim A, Alshawy F, Hussein C. Pencil cardinal Epigonus denticulatus Dieuzeide 1950, a western Mediterranean fish: Newly recorded from the Syrian marine waters (Eastern Mediterranean). Species 2023;**24**: e25s1025

[39] Saad A. Biology and Life Cycles of the Small Pelagic Fish on the Coast of Syria, Landing and the Catch Profile for the Syrian Coastal Fleet. TCP/SYR/4552. Rome: FAO; 1996. p. 45

[40] Soliman A, Ali M, Saad A, Renaud C, Capape C. First records of sideburn wrasse *Pteragogus pelycus* (Osteichthyes: Labridae) off the Syrian coast (eastern Mediterranean). Annals for Istrian and Mediterranean Studies, Series Historia Naturalis. 2014;**24**(1):23-28

[41] Khrema L, Saad A, Nesser A, Capape C. First Record of Baillon's Wrasse Symphodus bailloni (Labridae) in the Eastern Mediterranean Sea coast of Syria. FishTaxa. Turkey: Nevsehir Haci Bektas Veli University. 2022;**25**: 26-30

[42] Ibrahim A, Hussein C, Alshawy F, Alcheikh Ahmad A. First record of Pope's pony fish Equulites popei, (Osteichthyes: Leiognathidae) in the Syrian marine waters. Journal of Wildlife and Biodiversity volume (Special issue): (2020), pp. 2-4. DOI: 10.22120/jwb.20 20.123579.1127

[43] Alshawy F, Lahlah M, Hussein C. First record of the Berber ponyfish Leiognathus berbis Valenciennes, 1835 (Osteichthyes: Leiognathidae) from Syrian marine waters (eastern Mediterranean). Journal of Marine Biodiversity Records. 2016;**9**(98):2-4

[44] Ali M, Hammoud V, Fandi O, Capapé C. First substantiated record of crested oarfish Lophotus lacepede (osteichthyes: Lophotidae) from the Syrian coast (eastern Mediterranean Sea). In: Scientific and Research Center of the Republic of Sloveni. Vol. 31. No. 2. Annales: Series Historia Naturalis. 2021. pp. 205-210

[45] Saad A, Khrema L, Alnesser A, Barakat I, Capape C. First Substantiated Record of Dory Snapper *Lutjanus fulviflamma* (Forsskål, 1775) in the Eastern Mediterranean. Thalssia Sal. Italy: Marine Biology Station of Porto Cesareo; 2022;**44**:103-106

[46] Saad A. Biodiversité de la faune ichtyologique dans les eaux côtières de Syrie. In: la Méditerranée au 2I'eme siècle? variabilité climatique, environnement et biodiversité. Montpellier-France: Acte du colloque scientifique OKEANOS; 1995. pp. 185-189

[47] Ali M, Diatta Y, Alkusairy H, Saad A, Capapé C. First record of Red Sea goatfish Parupeneus forsskali (Osteichthyes: Mullidae) from the Syrian coast (eastern Mediterranean). Journal of Ichthyology. 2016b;**56**(4):616-619

[48] Sabour W, Masri M. First record of the Rosy goatfish, Parupeneus rubescens (Lacepede, 1801) (Mullidae) in Syrian marine waters. Iraqi Journal of Aquaculture. 2022;**19**(2):191-198

[49] Ali M, Saad A, Reynaud C, Capape C. First records of randall's threadfin bream *Nemipterus randalli* (Osteichthyes: Nemipteridae) off the Syrian coast (Eastern Mediterranean) off the Syrian coast. Annals for Istrian and Mediterranean Mar. Science. 2013; **19/2**:388-393 2018, 392 terranean Studies, Series Historia Naturalis, (2), 119-124

[50] Alshawy F, Ibrahim A, Hussein C, Lahlah M. New distribution of the serpent eel Ophisurus serpens (Linnaeus, 1758) in eastern Mediterranean: First record from the Syrian marine waters. International Journal of Agriculture & Environmental Science (SSRG-IJAES). 2019;**6**(3):50-52. DOI: 10.1444/23942568/IJAES-V6I3P110

[51] Capapé C, Saad A, Solaiman A, Barakat I, Sabour W. First Substantiated Record of Armless Snake Eel Dalophis imberbis (Osteichthyes: Ophichthidae) From the Syrian Coast (Eastern Mediterranean Sea). In: Scientific and Research Center of the Republic of Slovenia. Vol. 31. No. 1. Annales: Series Historia Naturalis; 2021. pp. 101-106

[52] Othman R, Galiya M, Almajid Z, Ghanem W. New record of the Roche's snake blenny Ophidion rochei Müller, 1845 (Ophidiidae, Ophidiiformes) from the Syrian marine waters – Eastern Mediterranean Sea. International Journal of Scientific Research in Biological Sciences. 2020;**7**(1):49-50

[53] Ali M, Saad A, Soliman A. Expansion confirmation of the indo-Pacific catfish, Plotosus lineatus (Thunberg, 1787),

(Siluriformes: Plotosidae) into Syrian marine waters. American Journal of Biology and Life Sciences. 2015c;**3**(1):7-11

[54] Saad A, Al Ksairy H, Sabour W. First record of the emperor angelfish, Pomacanthus imperator(Acthenoptergii: Pomacanthidae) in the Syrian coast (eastern Mediterranean). Marine Biodiversity Records. 2018;**11**(1):1-4. DOI: 10.1186/s41200-018-0152-3

[55] Capapé C, Hammoud V, Fandi A, Malek A.LI. First record of yellowbar angelfish Pomacanthus maculosus (Pomacanthidae) from the Syrian coast (eastern Mediterranean Sea). Acta Adriatica. 2023;**64**(1):93-97

[56] Saad A, Soliman A, Alkusairy H. First record of Abudefduf vaigiensis (Quoy and Gaimard, 1825) (Teleostei: Pomacentridae) in the Syrian coasts (eastern Mediterranean). Syrian Journal of Agricultural Research. 2020;7(4):478-485

[57] Capapé C, Hammoud V, Fandi A, Ali M. First Record of Moontail Bullseye Priacanthus hamrur (Osteichthyes, Priacanthidae) From the Syrian Coast (Eastern Mediterranean Sea). In: Scientific and Research Center of the Republic of Slovenia. Vol. 32. No. 2. Annales: Series Historia Naturalis; 2022. pp. 317-322

[58] Alshawy F, Ibrahim A, Hussein C, Lahlah M. First record of arrow bulleye, Priacanthus sagittarius Starnes, 1988 from the Syrian marine waters (eastern Mediterranean). FISHTAXA. 2019;**4**(2): 21-24

[59] Soliman A, Saad A, Ali M. First record of the blue-barred parrotfish, *Scarus ghobban* (Actinopterygii: Scaridae) from Syrian marine waters. Tishreen University Journal of for Research. and Science Studies. 2018; **40**(2):57-66

[60] Othman RM, Galiya MY, Laika HE, ALmajid ZA. First record of the lessepsian migrant mackerel, Scomber indicus,(Scombridae) off the Syrian coasts, Eastern Mediterranean. Acta Biologica Turcica. 2023;**36**(2):1-1

[61] Ali M, Alkusairy H, Saad A, Reynaud C, Capapé C. Confirmed occurrence of common lionfish, Pterois miles (Osteichthyes: Scorpaenidae) in the eastern Mediterranean, with first record off the Syrian coast. Tishreen University Journal for Research and Scientific Studies – Biological Sciences Series. 2016;**38**(4):307-313

[62] Fandi A, Hammoud V, Zeini A, Arraj T. First record of the Indo-Pacific red lionfish Pterois volitans (Linnaeus, 1758) in the Syrian marine waters (Banias). Journal of Hama University. 2022;5(19): 2022

[63] Al MS, Giovos I, Tiralongo F. New record of epinephelus areolatus in the mediterranean sea; first record from Syria. Annals Series Historia Naturalis. 2021;**31**(1):23-28. DOI: 10.19233/ashn.2021.04

[64] Saad A, Soliman A, AlKusairy H. First documented record of dogtooth grouper Epinephelus caninus (Valenciennes, 1834), Perciformes, Serranidae, in the Syrian marine waters. Syrian Journal of Agricultural Research – SJAR. 2020;7(1):465-470

[65] Foulquie M, Dupuy de la Grandrive R. First Assignment Concerning the development of Marine Protected Area on the Syrian coast, from 8-15 November 2002. RAC/SPA. 2003. p. 33.

[66] Ibrahim A, Lahlah M, Kassab MK, Ghanem W, Ogaily S. Signatus javus, a new record from the Syrian waters, with a reference to growth and feeding of two

Lessepsian fish. Rapport Commission international Mer Méditerranée. 2010; **39**:544

[67] Saad A, Masri M, Tufahha A. First substantiated record of the towbar seabream, Acanthopagrus bifasciatus (TELEOSTEI: Sparidae), in the Syrian marine waters (eastern Mediterranean Sea). Asian Journal of Advances in Research. 2022;**17**(3):14-18

[68] Sbaihi M, Saad A. Données nouvelles sure des espèces de poissons téleostéens pêchées pour la première fois dans les eaux terretoriales Syriennes. Proceeding des travaux de la semaine de science, Damas. 1992;**3**:83-105 (in arabic with abstract in french)

[69] Saad A, Masri M, Sabour W. b. Firstconfirmed record of Sparid *Pagellus bogaraveo* (Brünnich, 1768) in the Syrian marine waters (Levantine Basin). Marine Biodiversity Record. 2020;**13**(1): 1-7. DOI: 10.1186/s41200-020-0185-2

[70] Saad A, Sabour W, Masri M, Barakat I, Capape C. On the occurrence of red seabream *Pagrus major* (Osteichthyes: Sparidae) in the eastern Mediterranean Sea, first record from the Syrian coast. Cahiers de Biologie Marine. 2022;**63**:89-92. DOI: 10.21411/CBM.A. BEA5E160

[71] Hamwi N, Ali-Basha N. First record of Goldlined seabream Rhabdosargus sarba (Forsskål 1775), Sparidae, in the Mediterranean Sea (Syrian waters). Marine Biodiversity Records. 2021;**14** (12):1-5. DOI: 10.1186/s41200-021-00207-7

[72] Ibrahim A, Alshawy F, Hussein C. Stonefish *Synanceia verrucosa* Bloch & Schneider, 1801 (Actinopterygii: Synanceiidae): The first record in the Syrian coast and the fourth in the Mediterranean. International Journal of Aquatic Biology. 2019;7(6):383-386. DOI: 10.22034/ijab.v7i6.750

[73] Alshawy F, Ibrahim A, Hussein C, Lahlah M. c. First record of the oceanic puffer Lagocephalus lagocephalus (Linnaeus, 1758) from the Syrian marine waters (eastern Mediterranean). Marine Biodiversity Records. 2019;**12**(11):1-5. DOI: 10.1186/s41200-019-0170-9

[74] Khalaf G, Saad A, Gemaa S, Sabour W, Lteif M, et al. Population structure and sexual maturity of the pufferfish Lagocephalus sceleratus (Osteichthyes, Tetraodontidae) in the Lebanese and Syrian marine waters (eastern Mediterranean). Journal of Earth Science and Engineering. 2014; 7(4):56

[75] Sabour W, Saad A, Jawad L. First record of yellow spotted puffer Torquigener flavimaculosus Hardy & Randall, 1983 (Osteichthys: Tetraodontidae) from Mediterranean Sea coasts of Syria. Thalassia Salentina. 2014;**36**:29-34

[76] Ali M, Saad A, Ben Amor M, Capape C. First records of the honeycomb stingray, Himantura uarnak (Forskal, 1775), off the Syrian coast (eastern Mediterranean), (Chondrichthyes: Dasyatidae). Zoology in the Middle East. 2010;**49**:104-106

[77] Saad A, Ali-Basha N, Tufahha A, Barakat I, Capape C. First substantiated record of leopard whipray, Himantura leoparda (Myliobatoidei: Dasyatidae) from the Syrian coast (eastern Mediterranean Sea). FishTaxa. 2021;**19**:5-8

[78] Alkusairy H, Saad A. First record of Leucoraja circularis (Chondrichthyes: Rajidae) in the Syrian marine waters (eastern Mediterranean). Marine

Biodiversity Records. 2018;**11**(5):1-5. DOI: 10.1186/s41200-018-0140-7

[79] Saad A, Alkussairy H. First record of shagreen ray Leucoraja fullonica (Linnaeus, 1758) in Syrian coastal waters (eastern Mediterranean). Cahiers de Biologie Marine. 2019;**60**(3): 303-306. DOI: 10.21411/CBM.A.32012CD2

[80] Saad A, Barakat I, Masri M, Sabour W. First substantiated record of sea lamprey Petromyzon marinus (Agnatha: Petromyzonidae) from the Syrian coast (eastern Mediterranean Sea). FishTaxa. 2021;**20**:21-24

[81] MEA (Millennium Ecosystem Assessment). Ecosystemsand Human Wellbeing: Biodiversity Synthesis. Washington, DC: World Resources Institute; 2005. p. 86

[82] Galil BS. Alien species in the Mediterranean Sea—Which, when, where, why? In: Challenges to Marine Ecosystems. Developments in Hydrobiology. Vol. 202. Dordrecht: Springer; 2008. DOI: 10.1007/978-1-4020-8808-7_10

[83] Saad A, Sabour W, Soliman A. Contribution to the study fishing effort yeil of artisanal fishing gears: Quantitative and qualitative of catch composition in the marine waters of Tartous. Tishreen University Journal for Research and Science Studies Biology Sciences Series. 2016;**38**(1): 124-140

[84] Saad AA, Sabour WG. Impact of invasive species on the biodiversity and fish stock; case study: *Fistularia commersonii* Ruppel, in the Syrian coast. Rapports et procés-verbaux des réunions Commission internationale pour l'exploration scientifique de la mer Méditerranée. 2010;**39**:652

[85] The director of the National Hospital in Lattakia, Dr. Siham Makhoul, told Al-Khobar TV: "The First Case Received due to Poisoning from a Balloon Fish in Lattakia was in 2005". Available from: https://www.facebook.com/ScoopSyria/posts/117688493225248/

[86] Death of a Man in Baniyas, after they ate this Poisonous Buffer Fish. 2020. Available from: https://elissarnews.org/single?Id=20210106034203

[87] The Death of a Child in Lattakia on Tuesday 5/01/2021. Available from: https://www.syria.news/ab39d1ac-05012112.html (online newspaper)

[88] Saad A, Ghyath M. ATLAS of Toxic and Harmful Fish in the Syrian Marine Waters: Taxonomic Position, the Type of Harm Caused and How Treatment and Prevention. Damascus, Syria: Edit. Ministry of Higher Education; 2021. p. 112. Available from: https://www.researchgate.net/publication/348335317

Chapter 6

Fish Otolith Microchemistry as a Biomarker of Metal Pollution in the Estuarine Ecosystem

Abhijit Mallik, Suchismita Prusty, Puja Chakraborty, Shyamal Chandra Sukla Das and Shashi Bhushan

Abstract

Numerous metal pollutants naturally find their way into estuaries, where many of them build up in the bodies of fish. While otoliths can give a historical record of pollution exposure, metal concentrations in soft tissue and water samples require ongoing, long-term sampling procedures. Fish have otoliths, which are three pairs of ear bones called the sagitta, lapillus, and asteriscus. The chemical makeup of these otoliths can be a useful tool to determine the presence of hazardous substances in fish because the physiological activity of fish is controlled by a variety of environmental factors. The possible use of otoliths as inorganic tracers of metal contamination will be covered in this chapter.

Keywords: otolith, estuaries, pollution, biomarker, fish

1. Introduction

In aquatic environments, heavy metals stand out as the primary pollutants due to their widespread distribution and substantial presence. They tend to amass within the tissues of freshwater organisms, leading to fish poisoning and causing diverse effects across multiple biological levels, often resulting in pathological alterations. The majority of fish have otoliths, or "ear-stones," which are calcium carbonate structures that aid in hearing and balance and are located below the brain. The otoliths float in fluid-filled sacs in the inner ear and are not connected to the skull or any other bone. The three pairs of otoliths most common to fish are the sagitta (the largest pair), lapillus, and asteriscus [1]. Only fish with cartilage (such as sharks, rays, and chimeras) and fish without jaws (such as lampreys and hagfish) lack otoliths. Black drum (*Pogonias cromis*) and Atlantic croaker (*Micropogonias undulatus*), two fish that inhabit artificial habitats and vocalize during mating, have exceptionally large otoliths. The relatively smaller otoliths of species that swim continually, including bluefin tuna (*Thunnus thynnus*) and alewives (*Alosa pseudoharengus*), may be an attempt to control excessive movement. The most popular applications for sagittal otoliths, which are hard, porous calcium carbonate structures in teleost fish inner ears, are population and microchemistry investigations. They play a role in fish hearing, motion, displacement, sensation, and linear acceleration [2]. Calcium carbonate crystallites implanted in a protein lattice mineralize to produce otoliths, which then continue to grow [2]. Metals are among the major chemical toxicants polluting the environment due to their prolonged persistence and complex interactions with organisms in aquatic ecosystems. Consequently, changes in metal levels can be reflected in aquatic organisms, serving as biological indicators of metal exposure [3]. The concentration of metals in fish tissues mirrors their exposure through water and/or food and serves as a means to evaluate the present condition of these organisms before the onset of adverse toxic effects. Nevertheless, excretion mechanisms modify the accumulated metal burden over time. As assessments of accumulation levels are typically sporadic, they cannot serve as markers for constructing comprehensive lifetime accumulation histories. However, the toxic impact of metals is notably pronounced during the early phases of fish development, significantly affecting various metabolic processes and leading to developmental delays, morphological and functional abnormalities, or mortality in the most vulnerable individuals. Substantial accumulation levels throughout an entire lifespan or during critical developmental stages can detrimentally impact the future of populations. If necessary, trace elements can be absorbed into either the organic or inorganic component of the developing otolith. The elemental incorporation of metal contaminants in the aragonite matrix of the otolith is not a simple function as the process is species-specific [3], driven by environmental conditions [4], and route of exposure [5]. Studies on the application of otoliths in the metal exposure assessment of aquatic ecosystems and their comparison to soft tissues as bioindicators are rare. Thus, it is very important to assess whether metal accumulation in hard tissues (scales, otoliths) reflects metal exposure in correspondence to soft tissues (liver, muscle) of fishes [5].

2. Metal pollutants in estuaries

Estuarine ecosystems serve as crucial transition zones between ocean and river ecosystems, boasting high productivity in aquatic environments. These areas are home to a diverse array of fish species, many of which hold significant commercial value. Estuaries play essential ecological roles, serving as nursery areas for juvenile fish, feeding grounds for resident and adult fish, and temporary environments for the reproduction and migration of various species [6].

Given their ecological significance, estuaries provide valuable goods and services for human activities. However, these ecosystems face growing anthropogenic pressures, particularly from pollution. Estuaries naturally act as sinks for a variety of metals, accumulating them at concentrations that, while potentially highly toxic, are generally non-lethal. This accumulation of metals poses a threat to the overall health of estuarine ecosystems [7].

Metal pollution poses a worldwide challenge in estuaries, stemming from historical contamination legacies and ongoing increases in metal emissions. Nevertheless, establishing water and sediment standards or implementing effective management strategies in brackish systems has proven challenging, primarily due to the complex, multidisciplinary nature of estuarine processes. As per the European Commission, achieving an integrated understanding of the fate and effects of contaminants across various compartments of these transitional environments (including estuarine sediment, water, and biota) is still necessary to more effectively establish, evaluate, and monitor the targeted good ecological status outlined by the Water Framework

Directive [8]. Numerous environmental issues stemming from human activities like industrialization and urbanization are faced by this ecosystem. Physical, chemical, and biological processes all play significant roles in the metal biogeochemical cycle in this complex coastal environment. However, metal pollution in the coastal area has been brought on by anthropogenic factors, including urbanization and industrialization. Since most metals can poison organisms when present in amounts over threshold levels, this issue is getting more and more serious [9]. The sources include mining, the production of metal products, the disposal of solid waste, the burning of fossil fuels, and municipal and industrial waste effluent [10–12]. Particle size, amounts of particulate and dissolved organic matter, and many other variables all have an impact on the amount of exchangeable metal stored in the sediment. Thus, it indicated decreased cobalt and cadmium adsorption and enhanced zinc adsorption [13]. It is generally known that benthic fauna can serve as a sign of heavy metal contamination [14]. Several aquatic species including *Tittorina* sp., *Patella* sp., *Mytilus edulis, Scrobicularia plana, Macoma* sp., *Nucella* sp., *and Nereis* sp. are among the species and/or genus that are frequently employed. Heavy metal concentrations in all of these creatures could be thousands of times higher than they are in the water column. Therefore, heavy-metal concentrations can be easily and reliably determined as long as enough material can be gathered. However, seasonal fluctuations in metal content have been identified in several benthic mollusks [15]. The degree of toxicity of various metals in estuarine organisms varies depending on the chemical form of the metal in the solution [16]. The composition of both organic and inorganic ligands in the medium influences the free metal ion, which in turn affects the availability and toxicity of trace metals. As a result of complexation with chloride ions, it has been demonstrated that the concentration of free cadmium ions in solution varies inversely with salinity [17]. Other trace metals may behave similarly, making it likely that the estuarine biota's access to them will frequently rely on the surrounding medium's salinity. Unlike freshwater systems, where nutrients primarily limit primary production, the growth of phytoplankton in coastal zones is predominantly constrained by a complex interplay of factors, including nutrient loading, the system's filtering capacity, contaminants, light availability, and ecological interactions. For example, exposure to less than 90 $\mu g \cdot L^{-1}$ of Cu or Zn has been shown to result in growth reductions of up to 50% for cosmopolitan phytoplankton species. Laboratory experiments simulating metal exposure have indicated that phytoplankton may experience reduced cytokinesis, disturbances in photosynthesis, and increased cell size in response to various metals and species [8]. In contaminated estuaries, it is evident that cellular concentrations of metals may reach toxic levels, exerting control over growth and potentially leading to alterations in phytoplankton productivity and species composition [18]. Nevertheless, the quantitative significance of these effects under environmental conditions is yet to be fully elucidated.

Ultimately, the physical and chemical gradients within estuaries play a crucial role in shaping biogeochemistry and organism physiology, thereby influencing the exposure and toxicity of organisms. The significance of each gradient in determining the behavior and toxicity of metals can vary considerably and show seasonal variations. For example, the impact of salinity on partitioning coefficients may be significantly influenced by fluctuations in river flow, input of particulate and organic matter, and phytoplankton growth. Consequently, the behavior and effects of metals in estuaries are dynamic and depend on both environmental and biological factors. These dynamics must be taken into account when establishing site-specific water quality criteria and environmental quality targets [19].

3. Trace metal accumulation

To date, a comprehensive analysis has identified 51 elements in otoliths, encompassing major components such as calcium (Ca), carbon (C), oxygen (O), and nitrogen (N), as well as minor elements like sodium (Na), strontium (Sr), phosphorus (P), magnesium (Mg), potassium (K), chloride (Cl), and sulfur (S). Additionally, numerous other elements are present at trace levels. Otolith microchemistry has emerged as a sclerochronological biomarker for aquatic pollution by metals. The majority of these elements exist at micro and trace levels, necessitating thorough investigation for detection [20]. The incorporation of elements into the otolith can occur through one of three mechanisms: random entrapment in the crystal lattice, calcium substitution on the developing crystal surface, and binding to the organic matrix components. This suggests that the otolith's chemical composition reflects certain physicochemical features of the surrounding water. Recently, the otolith microchemistry of trace elements has gained recognition as a valuable technique for providing insights into a fish's life cycle and environmental history, particularly challenging information to obtain otherwise. However, its application has been limited in studies related to freshwater environments [21]. The concept of the bioaccumulation factor provides insight into the absorption of metals by fish from their environment. The variation in accumulation capacity in fish tissues is influenced by a combination of external and internal factors. The metal accumulation in fish tissues is species-specific, meaning that the concentration of certain metals may be higher in one tissue compared to another. Whereas the elemental signatures found in otoliths establish a baseline for quantifying pollution levels in the wetland [22].

The metal build-up has been studied using otolith chemical analysis. Many different trace metals, such as Sr., Mg, Mn, Fe, Cu, Co, Ni, Zn, Rb, Li, and rare earth elements, can be incorporated into the structure of carbonate minerals [23]. For many years, researchers have exploited the uptake of metals into the aragonite lattice of fish otoliths (ear bones) as a historical record of exposure to metals in polluted habitats. In the forensic chemical examination of crude oils, the relative abundance of two metals, in particular Ni and V, is utilized to help establish the origin of the oil. Ion size and crystallization temperature are two significant restrictions on trace metal substitution. The oxidation state of the elements, the alkalinity, and metabolism of the organism, the nutrition, and their quantities in the water could all affect ionic substitution in biogenic carbonates. The inner ears of teleost fish include calcified structures called otoliths. Otoliths are made up of calcium carbonate crystals, typically aragonite, that accumulate over the course of a person's lifetime within a protein matrix. The organic or inorganic part of the otolith can contain trace amounts of several different elements during creation. It is believed that the fish's environment has the biggest impact on the concentrations of these trace elements. According to their concentration in the surrounding water, certain elements, such as Sr., Ba, Hg, and Pb, seem to be deposited in otoliths [24]. Furthermore, changes in environmental contamination through time and space have been linked to the amounts of heavy metals in otoliths [25]. The element strontium, which frequently takes the place of calcium ions in carbonates, has by far received the most attention in relation to this phenomenon. Although a number of environmental and physiological factors influence how much strontium is incorporated into fish otoliths, the amount of strontium in the surrounding water is crucial. Zinc was found in some otoliths [26]; however, the precise distribution of zinc was unknown to them. The lithosphere, particularly the rocks encircling the water in which the fish lives, must be the primary source of zinc in the ecosystem. Zinc availability and uptake by fish, however, appears to be influenced by both the environment they are in and their physiological processes. Food appears to be the main source of zinc uptake in seawater. A similar analysis does not seem to have been done in freshwater. Additionally, the fish may very likely take up zinc by active ion transport through the gills. The heating regime, pH, alkalinity, and amount of particulate matter in the water are all dependent on the type of aquatic habitats, such as the availability of zinc, food-chain bio-magnification, and zinc flux. Analysis of zinc in addition to strontium may be able to provide temporally restricted information on habitat environment and fish migration if zinc exhibits a systematic distribution in otoliths along the life history transect [5].

4. Otolith microchemistry

Otolith microchemistry stands as a potent tool for investigating life stage dispersal, and regional connectivity, evaluating population structure, delineating estuarine nurseries, gauging connectivity between juvenile and adult populations and, to a lesser extent, serving as an indicator of environmental pollution. The continuous accretion of otoliths, coupled with their metabolic inertness and resistance to resorption, results in the permanent incorporation of various elements from the local environment into the crystalline matrix [27]. While otolith microchemistry can reflect a blend of local environmental chemistry and individual physiology, the resultant elemental composition acts as a distinctive fingerprint, functioning as a natural tag for distinguishing location and inferring ontogenetic changes, enabling the differentiation of otolith microstructures from various environmental conditions or locations. Consequently, marine organisms inhabiting impacted areas are likely to accumulate metals, including anthropogenic ones, from their surroundings and subsequently transfer or transport them into higher trophic levels of the food chain [28].

Otoliths, which are the calcium carbonate ($CaCO_3$) earstones of fish, offer a rich source of information about the life and environmental history of fish. However, the specific $CaCO_3$ polymorph forming the otolith represents a critical yet often overlooked aspect during otolith analysis. While otolith trace element chemistry data increasingly plays a crucial role in informing management decisions, recent research indicates that the $CaCO_3$ polymorphs—aragonite, vaterite, and calcite—can significantly influence the incorporation of trace elements in a non-trivial manner. While it has been traditionally believed that most fishes have otoliths composed of the aragonite $CaCO_3$ form, recent literature reports challenge this assumption, suggesting a more diverse landscape in otolith composition [29]. Studies on the microchemistry of otoliths in urban settings are particularly important since metal pollutants are more concentrated in urban areas than in natural ecosystems. Copper (Cu), zinc (Zn), and lead (Pb) are common metals found in metropolitan areas. Fish otoliths may include metals that have been deposited there from a variety of environmental compartments. They can be found in food sources, bonded to the sediment, present in sediment porewater, and also in the water column [30]. In order to estimate the possibility of a biological influence in aquatic environments, sediments are frequently evaluated as a line of evidence and known to be a repository of contaminants [31].

5. Conclusion

The fish otolith, commonly known as the earstone, has traditionally served as a timekeeping structure. However, recent years have seen a surge in interest in its use as a metabolically inert environmental recorder, driven in part by technological advancements. Various applications have emerged, including stock identification, determination of migration pathways, reconstruction of temperature and salinity history, age validation, detection of anadromy, utilization as a natural tag, and chemical mass marking. Some of these applications are challenging or even impossible to achieve using alternative techniques. Microsampling and the latest developments in beam-based probes now enable the coupling of elemental assays with daily or annual growth increments in the otolith, offering a detailed chronological record of the environment. Despite these advancements, there has been limited critical assessment of the assumptions underpinning environmental reconstructions or consideration of potential variations in elemental incorporation compared to other calcified structures. Drawing on insights from recent advances in geochemistry and paleoclimate research for comparison, it becomes evident that not all applications of otolith chemistry are firmly grounded, although some are poised to become powerful and perhaps routine tools for mainstream fish biologists.

Over the past three decades, otolith microchemistry has been a valuable tool in various studies, including those focused on the life history, migration patterns, and environmental ecology of commercially important fish stocks. Comparisons are made between the concentrations of elements and isotopes in otoliths and those in the water where the fish resides. The accumulation of heavy metals in fish otoliths is influenced by several factors, including the concentration of the specific heavy metal in the environment, its bioavailability, and the physiological state of the individual fish (which affects the exchange rate between external and internal environments). The chemical composition of otoliths is regulated by the physiological activity of fish, itself influenced by environmental conditions. The concentration of heavy metals in the calcified otoliths of fish serves as a tracer of environmental pollution exposures. Analyzing the microchemistry of fish otoliths provides a potent tool for monitoring pollution levels in aquatic ecosystems. These trace elements have the potential to serve as a biomonitoring tool, distinguishing between metals in different seasons. It is evident that the accumulation of heavy metals in fish otoliths depends on a variety of factors, including the concentration in the environment, bioavailability, the physiological state of the individual fish (affecting the exchange rate between the external and internal environments), the mechanisms of various species for detoxifying various metals, the growth rate of the individual fish (affecting the rate of accumulation of otolith material), and the affinity of certain heavy metals for particular fish species. In instances when metal pollution is low enough not to affect fish growth, it is more likely that the rate of metal accumulation will be higher since otoliths develop more quickly in individuals who grow more quickly. Other divalent cations, including Mg, Sr., Ba, Mn, CU, Zn, and Pb, as well as smaller monovalent cations like Li can alter the amounts of various elements across the otoliths of individual fish. However, it has also been proposed that bigger cations and anions, including Mg, can be integrated by being trapped within the crystal lattice as crystal inclusions.

Acknowledgements

The authors are thankful to ICAR-Central Institute of Fisheries Education, Mumbai, Inida-61, for providing all the necessary support to write this book chapter.

Conflict of interest

No conflict of interest.

Author details

Abhijit Mallik[1,2*], Suchismita Prusty[1], Puja Chakraborty[1], Shyamal Chandra Sukla Das[3] and Shashi Bhushan[1]

1 ICAR-Central Institute of Fisheries Education, Mumbai, India

2 Narayan Institute of Agricultural Sciences, Gopal Narayan Singh University, Jamuhar, Bihar, India

3 ICAR-Central Inland Fisheries Research Institute, Guwahati, India

*Address all correspondence to: abhijitmallik1993@gmail.com

References

[1] Aydin R, Calta M, Sen D, Coban MZ. Relationships between fish lengths and otolith length in the population of *Chondrostoma regium* (Heckel, 1843) inhabiting Keban Dam Lake. 2004;7:1550-1553

[2] Bäuerlein E. Growth and form: What is the aim of biomineralization? In: Handbook of Biomineralization: Biological Aspects and Structure Formation. Weinheim: Wiley-VCH Verlag GmbH & Co. KGaA; 2007. pp. 1-20

[3] Chang MY, Geffen AJ. Taxonomic and geographic influences on fish otolith microchemistry. Fish and Fisheries. 2013;**14**(4):458-492

[4] Campana SE. Chemistry and composition of fish otoliths: Pathways, mechanisms and applications. Marine Ecology Progress Series. 1999;**188**:263-297

[5] Ranaldi MM, Gagnon MM. Zinc incorporation in the otoliths of juvenile pink snapper (*Pagrus auratus* Forster): The influence of dietary versus waterborne sources. Journal of Experimental Marine Biology and Ecology. 2008;**360**(1):56-62

[6] Selleslagh J, Echard A, Pécheyran C, Baudrimont M, Lobry J, Daverat F. Can analysis of *Platichthys flesus* otoliths provide relevant data on historical metal pollution in estuaries? Experimental and in situ approaches. Science of the Total Environment. 2016;**557**:20-30

[7] Cappello T, Mauceri A, Corsaro C, Maisano M, Parrino V, Paro GL, et al. Impact of environmental pollution on caged mussels *Mytilus galloprovincialis* using NMR-based metabolomics. Marine Pollution Bulletin. 2013;**77**(1-2):132-139

[8] de Souza Machado AA, Spencer K, Kloas W, Toffolon M, Zarfl C. Metal fate and effects in estuaries: A review and conceptual model for better understanding of toxicity. Science of the Total Environment. 2016;**541**:268-281

[9] Prabhakaran P, Ashraf MA, Aqma WS. Microbial stress response to heavy metals in the environment. RSC Advances. 2016;**6**(111):109862-109877

[10] Fernandez-Luqueno F, Lopez-Valdez F, Gamero-Melo P, Luna-Suarez S, Aguilera-Gonzalez EN, Martínez AI, et al. Heavy metal pollution in drinking water-a global risk for human health: A review. African Journal of Environmental Science and Technology. 2013;**7**(7):567-584

[11] Engel DW, Fowler BA. Factors influencing cadmium accumulation and its toxicity to marine organisms. Environmental Health Perspectives. 1979;**28**:81-88

[12] Nemr AE, Khaled A, Sikaily AE. Distribution and statistical analysis of leachable and total heavy metals in the sediments of the Suez Gulf. Environmental Monitoring and Assessment. 2006;**118**(1):89-112

[13] Murray CN, Murray L. Adsorption-desorption equilibria of some radionuclides in sediment-fresh-water and sediment-seawater systems. Monaco: International Atomic Energy Agency; 1973. pp. 105-124

[14] Bryan GW. The occurrence and seasonal variation of trace metals in the scallops *Pecten maximus* (L.) and *Chlamys opercularis* (L.). Journal of the Marine Biological Association of the United Kingdom. 1973;**53**(1):145-166

[15] Bryan GW, Hummerstone LG. Indicators of heavy-metal contamination in the Looe estuary (Cornwall) with particular regard to silver and lead. Journal of the Marine Biological Association of the United Kingdom. 1977;**57**(1):75-92

[16] Tomlinson DL, Wilson JG, Harris CR, Jeffrey DW. Problems in the assessment of heavy-metal levels in estuaries and the formation of a pollution index. Helgoländer meeresuntersuchungen. 1980;**33**(1):566-575. DOI: https://doi.org/10.1007/BF02414780

[17] Sunda WG, Engel DW, Thuotte RM. Effect of chemical speciation on toxicity of cadmium to grass shrimp, *Palaemonetes pugio*: Importance of free cadmium ion. Environmental Science & Technology. 1978;**12**(4):409-413

[18] Dauvin JC. Effects of heavy metal contamination on the macrobenthic fauna in estuaries: The case of the Seine estuary. Marine Pollution Bulletin. 2008;**57**(1-5):160-169

[19] Pan K, Wang WX. Trace metal contamination in estuarine and coastal environments in China. Science of the Total Environment. 2012;**421**:3-16

[20] Leonhard I, Agiadi K. Addressing challenges in marine conservation with fish otoliths and their death assemblages. Geological Society, London, Special Publications. 2023;**529**(1):243-262

[21] de Almeida PRC, da Costa MR, de Oliveira RSC, Almeida A, Azevedo R, Monteiro-Neto C, et al. The use of the shape and chemistry of fish otoliths as a subpopulational discrimination tool for *Eugerres brasilianus* in lagoon systems in the Southwest Atlantic Ocean. Fisheries Research. 2023;**267**:106795

[22] Jenkins JA, Musgrove M, White SJO. Outlining potential biomarkers of exposure and effect to critical minerals: Nutritionally essential trace elements and the rare earth elements. Toxics. 2023;**11**(2):188

[23] Arai T, Ohji M, Hirata T. Trace metal deposition in teleost fish otolith as an environmental indicator. Water, Air, and Soil Pollution. 2007;**179**(1):255-263

[24] Milton DA, Chenery SR. Sources and uptake of trace metals in otoliths of juvenile barramundi (*Lates calcarifer*). Journal of Experimental Marine Biology and Ecology. 2001;**264**(1):47-65

[25] Dove SG, Gillanders BM, Kingsford MJ. An investigation of chronological differences in the deposition of trace metals in the otoliths of two temperate reef fishes. Journal of Experimental Marine Biology and Ecology. 1996;**205**(1-2):15-33

[26] Coote GE, Gauldie RW, West IF. Application of a nuclear microprobe to the study of fish otoliths and scales. Nuclear Instruments and Methods in Physics Research Section B: Beam Interactions with Materials and Atoms. 1991;**54**(1-3):144-150

[27] Rana A, Attri SV, Singh S, Kaur R. Otolith microchemistry of freshwater indigenous minor carp (Bangana dero) as a biomonitoring tool to trace heavy metals in River Sutlej, Ropar Wetland (Ramsar site), Punjab, India. Environmental Monitoring and Assessment. 2023;**195**(11):1286

[28] Vrdoljak D, Matić-Skoko S, Peharda M, Uvanović H, Markulin K, Mertz-Kraus R. Otolith fingerprints reveals potential pollution exposure of newly settled juvenile Sparus aurata. Marine Pollution Bulletin. 2020;**160**:111695

[29] Pracheil BM, George R, Chakoumakos BC. Significance of otolith calcium carbonate crystal structure diversity to microchemistry studies. Reviews in Fish Biology and Fisheries. 2019;**29**:569-588

[30] Andronis C, Evans NJ, McDonald BJ, Nice HE, Gagnon MM. Otolith microchemistry: Insights into bioavailable pollutants in a man-made, urban inlet. Marine Pollution Bulletin. 2017;**118**(1-2):382-387

[31] Simpson SD, Radford AN, Nedelec SL, Ferrari MC, Chivers DP, McCormick MI, et al. Anthropogenic noise increases fish mortality by predation. Nature Communications. 2016;**7**(1):1-7